Back to the Breast

Back to the Breast

Natural Motherhood and Breastfeeding in America

JESSICA L. MARTUCCI

THE UNIVERSITY OF CHICAGO PRESS CHICAGO AND LONDON

JESSICA L. MARTUCCI is a fellow in the Department of Medical Ethics and Health Policy, Perelman School of Medicine, University of Pennsylvania.

The University of Chicago Press, Chicago 60637
The University of Chicago Press, Ltd., London

Printed in the United States of America

24 23 22 21 20 19 18 17 16 15   1 2 3 4 5

ISBN-13: 978-0-226-28803-1 (cloth)
ISBN-13: 978-0-226-28817-8 (e-book)
DOI: 10.7208/chicago/9780226288178.001.0001

Library of Congress Cataloging-in-Publication Data

Martucci, Jessica L., author.
Back to the breast : natural motherhood and breastfeeding in America / Jessica L. Martucci.
pages ; cm
Includes bibliographical references and index.
ISBN 978-0-226-28803-1 (cloth : alk. paper) — ISBN 978-0-226-28817-8 (ebook)
1. Breastfeeding—United States—History—20th century. 2. Breastfeeding promotion—United States—History—20th century. 3. Infants—Nutrition—United States. 4. Motherhood—United States. 5. Maternal and infant welfare—United States—History—20th century. 6. Mothers—United States—Social life and customs—20th century. I. Title.
RJ216.M36 2015
649′.33—dc23

2015004416

Portions of the introduction and chapter 3 first appeared in *Journal of Women's History*, © The Johns Hopkins University Press. An earlier version of chapter 4 appeared in Jessica Martucci, "Maternal Expectations: New Mothers, Nurses, and Breastfeeding," *Nursing History Review* 20 (2012): 72–192. Springer Publishing Company, LLC.

♾ This paper meets the requirements of ANSI/NISO Z39.48-1992 (Permanence of Paper).

Contents

Introduction

Why Breastfeeding?

In 1956, Martha, a mother of four and advocate for maternal health in Cleveland, Ohio, wrote to the Maternity Center Association headquartered in New York City. Her letter addressed an issue that had long been on her mind: breastfeeding. In the 1940s and 1950s, Martha belonged to a shrinking cohort of American mothers who breastfed their infants for any length of time. As a member of the Association for Parent Education group in her area she knew of a wide range of information and resources on natural childbirth and breastfeeding. Unable to avoid the "black out" anesthetic options imposed on her during the labors and deliveries of her first two children in the hospital, she was all the more happy when she was able to counteract these unwanted interventions by successfully breastfeeding all four of her children. She fondly reminisced in her letter that she had "weaned our last baby just one week before her first birthday and enjoyed nursing her even more than our first three." Despite her own personal breastfeeding triumphs, she lamented what she believed was the growing need for a network of maternal advocates who might support mothers who nursed. "For breast feeding there is no product, thus no sponsor," she observed, and yet she knew many mothers who wanted to breastfeed. "I have help [*sic*] 15 mothers myself . . . but feel that a larger group organized for this purpose would be more effective to a larger group of women."[1] Little did Martha know at the time that a handful of like-minded mothers in Franklin Park, Illinois, were already in the process of organizing just such a group. Martha's interest in breastfeeding eventually led her to discover La Leche League's existence and she went on to found the nation's second League chapter.

The history of infant feeding has long held an important place in the interdisciplinary body of scholarship devoted to gender, sexuality, women's studies, and the history of medicine.[2] Martha's story, however, hints at an important narrative thread in the history of breastfeeding that has remained underexplored: the persistence of breastfeeding throughout the age of the bottle and its impact on infant feeding trends, practices, and policies later in the century. How do we make sense of mothers like Martha who remained dedicated to breastfeeding throughout this period? In the 1950s, bottle-feeding, not breastfeeding, was widely supported by medical advice as well as the normative ideals of motherhood, science, and consumerism that characterized the postwar era.[3] Yet Martha was not alone in her devotion to breastfeeding in this period, a fact evidenced by the breastfeeding angst in the popular press and the success of women's health education networks, of which La Leche League was only one part. Women like Martha cultivated, manipulated, and spread knowledge about breastfeeding among themselves while relying on a small but prolific group of midcentury doctors, nurses, and social and biological scientists. Together, they created a body of expert knowledge about motherhood, lactation, and child rearing that provided legitimacy and validation for their perspective.

That the early roots of the breastfeeding movement can be found only in low-profile subcultures within American science and culture at the time should not lead us to automatically dismiss the importance of these seemingly trivial beginnings. If anything, the persistence of breastfeeding as a practice and a body of knowledge during the decades of bottle-feeding's vast popularity demands closer attention and analysis in order to better contextualize the story of breastfeeding's sharp rise in the last quarter of the twentieth century. That breastfeeding knowledge survived, and arguably thrived, throughout small pockets of American science and culture during the middle third of the last century is a reality that has not yet been adequately addressed by the existing historiography.[4] In this book, I trace the emergence, rise, and fraught continuation of breastfeeding in the twentieth century into the twenty-first. What can we learn about motherhood throughout this period from examining the growth of a breastfeeding movement in the mid-twentieth century? What historical insights can we gain through an analysis of the uniquely modern connection between breastfeeding and motherhood that emerged in those decades? How does focusing on a story of breastfeeding's persistence change our understanding of its history? Who were the people involved in reframing breastfeeding and why did it hold value for them? How did changing ideas about science, nature, and motherhood influence what breastfeeding meant in the postwar period and how do the ideological roots of its renewal continue to shape the practices, meanings, discourse, and policies surrounding breastfeeding today?

In pursuing these questions, this book explores the circumstances, ideas, actions, and legacies of the mid- to late twentieth-century mothers, scientists, and clinicians, who advocated, wrote about, studied, and practiced breastfeeding. The history behind breastfeeding's return reveals an important intersection between the experiential knowledge of mothers and the scientific expertise of professionals in the medical and human sciences. My focus in this project is on the ideas, practices, and processes surrounding breastfeeding. I do not engage substantively with narratives in the history of formula feeding, the baby-food industry, or the activism surrounding these issues (i.e., the Nestlé boycott). By concentrating explicitly on breastfeeding, this account provides readers with a more complex and nuanced perspective of the story behind the statistical rise in breastfeeding that began in the 1970s and offers a deeper historical context for contemporary breastfeeding discourse. Mothers and experts alike made breastfeeding matter in new ways in the mid-twentieth century by constructing an alternative ideological framework that I refer to as natural motherhood, one that relied overtly on a scientifically validated understanding of breastfeeding as an inherently "natural," "pure," and evolutionarily perfected and embodied process.[5] This ideology, which I discuss at length in the pages ahead, began to come together in the interwar years, but it did not completely coalesce until the postwar decades when it met with the era's pronatalism within the crucible of the Cold War family.[6]

The period encompassing World War II and the postwar years gave rise to a unique construction of breastfeeding within a new understanding of nature, one built around a science of instinct, evolutionary principles, and an evolving consciousness of the relationship between the natural world and that of humans, particularly women. This ideological underpinning helped breathe new life into breastfeeding by the 1950s, a phenomenon best illustrated in many ways by the founding of the breastfeeding support organization La Leche League in 1956, which continues to operate to this day as an international organization. Interest in breastfeeding slowly gathered momentum throughout much of the next two decades as it intersected with the burgeoning environmental and feminist movements of the 1960s and 1970s. For mothers in these years who sought to embrace both a feminist existence and a "back to the land" ethos popular during these tumultuous decades, natural motherhood ideology offered a model of breastfeeding that held widespread appeal.

The connection between natural motherhood and breastfeeding reforged in the postwar era took on new and important meanings in the years following the publication of Rachel Carson's *Silent Spring* (1962). In the years following Carson's work, which exposed the dangerous consequences of chemical pesticides on human and animal health, mothers became increasingly aware of the problem of toxins in their bodies and their breast milk. Though researchers had long known about the ability of DDT to pass through mother's milk, it took several decades to build an environmental movement aimed at regulating industrial and agricultural pollution in the name of human health. Along the way, mothers came together to protect the environments in which their children lived as well as the nature within their own bodies. Beginning with the work of Carson, mothers who breastfed in the second half of the twentieth century struggled to create a meaningful maternal experience based on their inescapable knowledge of polluted bodies and breast milk.

Furthermore, in light of the activism of second-wave feminists during the same decades, connections between breastfeeding and natural motherhood ideology symbolized both a path to female pride and empowerment as well as an Achilles' heel. By the end of the 1970s and the beginning of the 1980s, postwar constructions of natural motherhood and breastfeeding appeared increasingly untenable as feminist allies. Arguments of "the natural" in these years more often than not surfaced in efforts to combat feminist economic and political gains. In this climate, popular constructions of breastfeeding as a feminist cause seemed suspect at best, and "old-fashioned" and traditionalist at worst. Despite the public turn in breastfeeding discourse, this period witnessed a dramatic rise in breastfeeding rates among white, middle-class mothers: an irony best explained, perhaps, by the continuing connection to "the natural" that breastfeeding offered. The experience of breastfeeding remained an intimate and deeply personal one regardless of the broader political overtones. At the end of the twentieth century, an increasingly reductionist construct of biological motherhood and the accumulating science of breastfeeding surfaced time and time again as part of a largely conservative movement to critique women's place outside of the home through the mobilization of "mother guilt."[7]

Breast pump technology assumed unprecedented importance during this period due to the rising popular and medical interest in breastfeeding in an era when more women were working outside of the home through their childbearing years than ever before.[8] A technology with a long and fascinating history in its own right, the breast pump moved into the spotlight of breastfeeding debates by the early 1990s. Hailed by many as the technological fix that could sever breastfeeding from its ideological baggage, the pump offered to ease the guilt of overburdened working mothers and even held the promise of expanding access to breastfeeding beyond the white middle class. The pump, of course, did change things but not always in the ways in which many hoped or imagined. By the end of the first decade of the twenty-first century, this device had become embedded within the process of breastfeeding to such an extent that the phrase "I'm breastfeeding" more than likely meant "I'm breast pumping." This dramatic and relatively quick shift in the breastfeeding process had equally dramatic consequences on the late twentieth-century emergence of a new health profession, that of the lactation consultant. Today "breastfeeding" often means any process by which an infant receives breast milk and it is now the purview, not primarily of pediatricians or even of mothers or women's lay groups, but of the certified lactation consultant.

Despite the relatively recent emphasis on technology in breastfeeding practice, many mothers today continue to embrace breastfeeding in ways that resonate with the midcentury's ideology of natural motherhood. Today, there is a notable discrepancy between the persisting relevance of natural motherhood for many women who breastfeed and the rise of an increasingly medicalized and bureaucratized system of lactation support. The vision of motherhood and breastfeeding articulated by Martha in 1956 had very little to do with health care professionals, hospitals, breast pumps, or the government. In tracing the story behind the rise of breastfeeding rates over the last half century, this book seeks to understand how and why the relationship between natural motherhood and breastfeeding evolved and changed. It is my hope that in doing so, we might come to better understand what is at stake when we talk about, argue over, and make policies about breastfeeding.

In writing this account on what continues to be a controversial topic in American society, it is also my hope that a more general readership will find this an engaging, provocative, and informative read. Like much of the history of family and motherhood, I would argue that the history of breastfeeding has the potential to spark interest in a readership far beyond the academy. The intimate details and rich experiences that histories of the body and of the home are able to offer, I believe, are inherently interesting to us because we are often able to see direct connections to our own lives. In order to facilitate readings by both academic and nonacademic readers alike, then, I have inserted a purposeful break here. For those who wish to understand my theoretical and methodological interventions into the thriving literature on the history of motherhood, medicine, and infant feeding, the remainder of the introduction will hopefully satisfy. For those who are not as interested in historiography, I recommend moving ahead to chapter 1, where the narrative picks up.

Organization

The chapters that follow examine how breastfeeding became rooted in natural motherhood. Through an exploration of the lives and work of scientists, medical researchers, physicians, nurses, lay groups, and mothers themselves, this book provides an account of the resurgence of breastfeeding practice and traces its shifting meanings since the 1930s. The book is organized into three chronologically overlapping yet thematically distinct parts. In the first two chapters, I explore the role that scientific and medical expertise played in the early history of breastfeeding's return over the middle third of the twentieth century. Chapter 1 establishes the history of the emergence of natural motherhood in the "psy-ences" in the 1930s, 1940s, and 1950s. The midcentury experts who helped reconstruct breastfeeding within the ideology of natural motherhood have occasionally received mention in historical works, but for the most part these contributions have remained underanalyzed. The interdisciplinary Columbia University–trained psychologist Niles Newton, for example, began publishing her work on the connections between emotions and physiological processes in the maternal body in the 1940s.[9] Despite her success, however, Newton remained at the periphery of her field and has been largely absent from discussions on the history of breastfeeding. Others in the midcentury sciences who devoted themselves to what I refer to as a woman-centric or maternal-centric perspective in the study of motherhood tended to inhabit temporary, marginal, and interdisciplinary positions due to their own social positions as "outsiders," often because of their gender. Chapter 1 looks closely at this group of experts who supported breastfeeding (sometimes consciously, sometimes not) through their research, their popular work, and their own clinical and parenting experiences. When we take their work into account alongside that of their more mainstream peers, we learn that the results of research into maternal and infant behavior and biology could and did hold different meanings depending on where one stood at the time.

The second chapter explores the persistence of an interest in breastfeeding within the medical world from the early twentieth century through the 1960s. While historians have thoroughly examined the twentieth-century rise of a medical paradigm that supported bottle-feeding and undermined breastfeeding, scholars of this subject have spent far less time analyzing the perspectives and efforts of those few doctors who continued to actively embrace the study and practice of breastfeeding throughout this time.[10] By focusing on the work and knowledge of doctors who supported breastfeeding in the age of the bottle, I move beyond existing accounts in the historiography of breastfeeding. The irony revealed by this analysis is that even those few who continued to study and advocate breastfeeding all too often failed to manage it successfully in their own practices. To explain the disconnect between the desire of these physicians to promote breastfeeding with their inability to effectively do so, I suggest that the midcentury emergence of breastfeeding as part of the ideology of natural motherhood made it a practice that was both inscrutable to medical expertise and resistant to the intervention of masculine authority and technology.[11] In short, breastfeeding success depended upon an ongoing relationship between a mother and her caregiver built on patience, trust, and support, a kind of patient-doctor relationship that was unlikely to occur in the wake of increasing specialization within medicine in the mid-twentieth century. Furthermore, breastfeeding, even more so than childbirth, demanded individualized attention and an expansive network of tacit knowledge to address issues appropriately, things few doctors had access to as the century wore on.

The next two chapters examine the role that women as mothers and nurses played in the history of breastfeeding's survival throughout the midcentury. The third chapter examines the culture surrounding breastfeeding mothers in the postwar years in order to provide a better understanding of the degree to which the ideology of natural motherhood infiltrated popular discourse and influenced mothers' experiences with breastfeeding. This chapter also demonstrates how difficult it could be for women who made the choice to breastfeed in these years, not only because they challenged the medical system, but because they also threatened the order of the postwar family. The fourth chapter continues to explore the contributions and experiences of women professionals to the history of breastfeeding through an analysis of the roles that nurses played in the process throughout the postwar decades. Nurses remained loyal to and bounded by the mid-twentieth century's tenets of scientific motherhood to such an extent that even those who experienced breastfeeding as mothers themselves often could not find a way to help others breastfeed within the system in which they worked. Despite the vast majority of nurses in the postwar period who remained agents of medical authority over infant feeding, however, some learned to cultivate a new role for themselves as natural motherhood experts. These nurse-mothers offered a unique form of scientific expertise and tacit knowledge which allowed them to serve as intermediaries in the disconnect that mothers often experienced between their desire to breastfeed and a medical system that supported bottle-feeding.

The remaining chapters push the book's narrative up through the end of the twentieth century. Chapters 5 and 6 focus on the perspective and experiences of mothers, lay groups, and medical professionals from the mid-twentieth century through the 1990s, when breastfeeding had become established as a central component of a global public health message. Chapter 5 examines how women who came of age in the late 1950s, 1960s, and 1970s viewed breastfeeding in light of its reconstruction under the ideology of natural motherhood. Parallels and clashes between the breastfeeding movement and environmentalism and feminism contributed to a polarization in the breastfeeding community. Disagreements over what breastfeeding should be and whether the ideology of natural motherhood could (or should) fit within a modern feminist framework left it unattached to any unifying ideology and stripped it of much of its potential to challenge constructions of motherhood. In this era, feminist ambivalence over how to deal with breastfeeding ironically helped entrench natural motherhood and breastfeeding within modern conservative arguments in favor of "traditional" family values.

Chapter 6 tells the story of the rise of the breast pump as a definitive component of breastfeeding by the end of the twentieth century. In doing so, I explore how the construction of breastfeeding as the embodiment of natural motherhood was challenged by breast pump technology and the professionalization of breastfeeding expertise. The rise of the breast pump helped mothers begin to reframe breastfeeding in a more feminist light and played with the expectation of what "the natural" encompassed. In the epilogue, I offer an analysis of how the events of the previous century have continued to shape mothers' experiences with breastfeeding in the twenty-first century and suggest ways in which we might address the ongoing tensions surrounding the growing admonition that mothers should go "back to the breast."

The History of Breastfeeding

Breastfeeding as a practice slowly fell out of favor over the course of the late nineteenth and early twentieth centuries, when cow's milk, scientific formulas, and proprietary infant foods gradually replaced breastfeeding as the "normal" sources of infant nutrition. Between 1890 and 1950 infant feeding practices shifted dramatically alongside broad changes in understandings of health and disease; the influence and authority of the medical professions; women's domestic, economic, and political roles; and patterns of consumption and food production. All of these well-documented trends contributed to a rise in faith and interest in scientific methods of infant feeding alongside a decline in women's desire to nurse. Perhaps most importantly, this period also witnessed the dwindling faith by mothers and medical experts alike that the modern American woman even *could* breastfeed.[12] Historians have analyzed these trends alongside evidence of diminishing medical interest in breast milk to suggest that by the 1930s and 1940s breastfeeding was no longer an issue; mothers by that time simply accepted bottle-feeding as the norm.[13] An extensive body of interdisciplinary academic and popular literature explores this decline throughout Western industrialized nations during the last century.[14] These works have tended to overlook the significant number of mothers who continued to pursue breastfeeding and the peripheral network of scientists and experts who supported them throughout the mid-twentieth century, the period when breastfeeding rates were at their lowest.[15] Despite the continued decline in breastfeeding throughout the 1930s, 1940s, 1950s, and 1960s, these years simultaneously gave rise to a diverse and important scientific and lay discourse on motherhood and infant feeding, one with profound consequences for mothers at the end of the twentieth century.

Writing the history of a practice during a time when it was at a historical low point poses particular historiographical and methodological challenges. Works like Rima Apple's *Mothers and Medicine* offer well-documented narratives of just how little doctors or mothers alike seemed to concern themselves with breastfeeding in the postwar years. Drawing on popular advice and medical literature from the period 1890–1950, Apple argues that a significant decrease in the amount of public discourse on the breast versus bottle debate suggests that by the 1950s, bottle-feeding was no longer a choice but simply a reality.[16] In this narrative, in the late 1890s the emergence of a "coherent ideology" of scientific motherhood helped drive women away from traditional sources of child rearing and infant feeding knowledge toward the growing body of scientific experts who offered solutions to modern women's problems.[17] In turning to physicians for guidance, however, women ceded authority and control over their child-rearing practices, including infant feeding. As time progressed, the development of scientifically designed formulas facilitated medical oversight and control over infant health and contributed to a diminishing faith in the abilities of women's bodies to feed their babies.[18] At the same time, women themselves actively participated in a long tradition of seeking alternatives to breastfeeding, eventually trusting in science and technology to provide them with safe infant foods that alleviated some of the burdens of modern motherhood.[19] Coupled with the rise of unchecked relationships between commercial infant food companies and physicians, the medical establishment and mothers themselves helped construct bottle-feeding as "normal" over the first several decades of the twentieth century and the practice became institutionalized in the hospital's maternity and postnatal procedures.[20]

Subsequent works on the history of infant feeding and scientific motherhood in the American context by scholars such as Jacqueline H. Wolf and Janet Golden, while both broadening and deepening our understanding of the historical processes involved in the decline in breastfeeding, have supported this meta-narrative of the ideology of scientific motherhood and the ubiquitous embrace of formula feeding which arguably accompanied it.[21] Wolf has convincingly shown that mothers themselves played a central role in the normalization of bottle-feeding as women adopted scheduled feedings and bottles to facilitate the busy schedules of modern urban life.[22] Golden's work also expands our understanding through her work on wet nursing, which declined dramatically in the first two decades of the twentieth century, a trend she links, among other things, to the commodification of breast milk by the end of the 1920s.[23] The commodification of breast milk further supports the argument that by the middle of the century, breastfeeding had become unimportant and detached from any particular maternal identity, or body.[24]

Despite these important contributions, however, scholarly analysis of the persistence of breastfeeding in the doldrums of the mid-twentieth century remains elusive. Furthermore, there has been a tendency to assume that the trends in breastfeeding's decline, established by the beginning of World War II, remained in place, largely unchanged until breastfeeding became interesting again in the wake of second-wave feminism and a "back to the land" movement that encouraged living a "more natural" life.[25] Exceptions to this have been studies that have focused on the history of La Leche League. This body of work has contributed to a complex portrait of the League and its role in shaping the discourse and experience of breastfeeding in the second half of the twentieth century.[26] Formed in a wealthy Chicago suburb by a group of Catholic mothers in 1956, the League's very existence challenges assumptions that the postwar era contributed little to changes in the ideas and practices surrounding breastfeeding and motherhood. La Leche League's temporally radical message of natural and "better" motherhood through breastfeeding positioned it as an outsider in a culture largely devoted to Big Science, consumerism, technological innovation, and personal fulfillment through the pursuit and attainment of marriage, kids, and a house in the suburbs.[27] Analyses of La Leche League, however, tend to characterize the organization either as a conservative, and ultimately failed, attempt to reinstate a Victorian vision of motherhood, or just as limiting, as a false start to the feminism that would emerge in the 1960s and 1970s.[28] The most successful treatment of La Leche in its historical context remains Lynn Y. Weiner's 1994 article, "Reconstructing Motherhood: La Leche League in Postwar America," in which she suggests that the League arose "to defend traditional domesticity against the assaults of modern industrial life and to dignify the physical, biological side of motherhood."[29] According to Weiner, the League accomplished this through an emphasis on "naturalism." While providing innovative and insightful analysis, Weiner stops short of following the League's interest in "naturalism" to its full extent. Why did arguments built on "naturalism" hold particular sway in this period? From where did the knowledge about how and why natural breastfeeding was best come? Though she acknowledges that it was this appeal to "the natural" that made breastfeeding interesting to feminists in the 1970s, Weiner's analysis does not provide a clear map for understanding the challenges and changes that the midcentury breastfeeding movement faced when it encountered the social upheaval of the late 1960s and 1970s. Further, in labeling the League a "maternalist" organization, Weiner circumscribes the extent to which their ideas and ideologies reflected a much larger network of mothers and experts who participated in the construction of an alternative ideology of motherhood and breastfeeding.

Historical work on the League has also tended to portray it as an isolated organization. The life histories of its original founders and medical supporters, the speed of its expansion, its grassroots structure, its widespread influence on mothers and health care professionals via its publications and outreach, and its interactions with other women's groups all suggest that La Leche League emerged in response to broader trends already in motion by the time it officially came together in 1956.[30] Without scientific infrastructure and evidence to support and inform their ideas, without far-reaching interest both from within the scientific world and beyond, and without the participation of mothers and fathers from across the country, La Leche League would have simply remained a group of seven Catholic mothers outside of Chicago who occasionally got together to swap breastfeeding stories.[31] That it grew in size and influence relatively rapidly suggests that La Leche acted more as a platform for connecting and supporting people, ideas, and practices already in existence than as an isolated fringe group. Weiner's work, in particular, suggests that this persistence was connected in the postwar years to debates and anxieties over cultural constructions of motherhood, particularly the waning authority of women in the postwar home.

Focusing on those who maintained an interest in breastfeeding, mothers and experts alike, helps reveal an important but largely neglected story of continuity and of paths not taken. Low rates of breastfeeding unquestionably characterized the midcentury landscape of infant feeding, yet there are only a few studies from which our knowledge of this phenomenon is drawn. An oft-cited 1948 study by Katherine Bain, deputy chief of the Children's Bureau and a pediatrician, showed that 38 percent of mothers breastfed exclusively at the time of hospital discharge.[32] However, Bain's work also indicated that over two-thirds of the study participants had tried breastfeeding, and nearly as many continued to partially breastfeed at the time of hospital discharge.[33] A 1958 study of over seven hundred mothers from around Seattle revealed that 50 percent of the participants initiated breastfeeding.[34] While scientific motherhood and bottle-feeding continued to dominate in very real and measurable ways, the persistence of mothers who tried to breastfeed in spite of the widely condoned practice of formula feeding suggests the presence of a significant counter-ideology in support of breastfeeding. Additionally, those mothers who continued to breastfeed despite the many obstacles in their paths gained legitimacy and personal fortitude through their encounters with information gleaned from scientific and medical knowledge produced and popularized by the media and experts who favored breastfeeding.

The existence throughout the midcentury of a network of professionals and mothers who championed breastfeeding begs further explanation and analysis. The statistical reemergence of breastfeeding practice was necessarily preceded by the development of an ecological and evolutionary view of motherhood and nature, which I refer to as the ideology of natural motherhood. In cultivating this alternative ideology of "the natural," women and their supporters found theoretical grounding, encouragement, and validation in the work coming out of the human and animal sciences in the middle third of the twentieth century as well as in culturally resonant ideas about "the natural" world.

The Psy-ence behind Natural Motherhood

In the early decades of the twentieth century, breastfeeding rates fell and the popularity of alternative feeding methods rose. Experts on child rearing in this period focused their scrutiny on the physical health and hygiene of the infant, schooling mothers in the latest scientific practices for minimizing the spread of disease and optimizing their family's health through the application of nutritional science and technological innovations.[35] Efficiency, cleanliness, nutrition, and a belief in the healing properties of certain natural elements, particularly sunlight and seaside breezes, informed a new generation of parents and an expanding field of experts on child rearing.[36] In the century's first two decades, the psychological study of child development remained in its infancy and those who did explore the minds of children concentrated largely on those who found themselves in an evolving juvenile court and welfare system.[37] In this context, ideas of maternal competency mapped directly over the degree to which mothers learned, understood, and employed scientific methods in child rearing. Experts and lay people alike came to viciously denigrate any reliance on the knowledge and abilities of women on matters of infant care. To listen to the advice of grandmothers or other anecdotal evidence collected from neighbors and friends rapidly became suggestive of one's origins in a low and uneducated class. In this context, white, upper- and middle-class mothers led the movement away from the breast, followed by those with aspirations for higher status and eventually by those impacted by changing maternal and infant health policies.[38]

L. Emmett Holt, the renowned early twentieth-century pediatrician and author, was one of the most famous and widely read experts who denounced the existence of anything resembling "maternal instinct," and instead stressed that knowledge of balanced nutrition and the counsel of a trained physician in all child health matters were the key elements in successfully raising a child.[39] Infant feeding under Holt's method was routinized, sterile, and efficient, as were all mother-infant interactions. He stressed that infants should not be played with "under six months old" and that kissing was "under no circumstances" to be allowed unless "upon the cheek or forehead, but the less even of this, the better."[40] Holt's methods preceded the even more emotionally austere theories of John B. Watson. Watson's 1928 book *Psychological Care of Infant and Child* argued that mothers were responsible for the emotional health of their children, as well as the physical. Watson's writing attacked mother love to an even greater extent and contributed to the widespread belief that women could coddle their babies (especially sons) into useless neurotics. Early twentieth-century experts argued that while breast milk might be nutritionally superior and bacteriologically safer in most cases, the distance that bottle-feeding created between a mother and an infant was perhaps, psychologically, the safest alternative.[41]

Beginning in the 1930s, however, a new framework began to emerge through the interdisciplinary work of psychologists, psychiatrists, anthropologists and ethologists (an interdisciplinary and motley crew whom I refer to as "psy"-entists) which offered a challenge to the reigning ideology of scientific motherhood and the model of infant feeding it encouraged.[42] The ideology of natural motherhood reflected this set of scientific and cultural beliefs, which posited that women's bodies contained instinctual knowledge that could be unlocked through "natural" behaviors like childbirth and breastfeeding. Unlike scientific motherhood, which was dominated by a belief in the superiority of the knowledge and technologies of scientific experts, natural motherhood supported maternal authority over childbirth and infant care. Natural motherhood was not, however, antiscience. It was built on accepted scientific research surrounding maternal and infant behavior suggesting that technological meddling could irreparably harm the evolutionarily perfected, "natural," and presumably necessary connections between mother and child.[43] The 1940s and 1950s witnessed a series of high-profile scientific studies released in a cultural moment when anxieties over women's roles were exceptionally high.[44] And while this research can be easily viewed as part of the larger cultural backlash against mothers, scientists and mothers alike also saw in this work the grounds for a moral argument in favor of "natural" mothering. This scientific basis for the ideology of natural motherhood provided a framework in which women could argue for greater autonomy over maternal practices. The science of maternal instinct emboldened some women to seek purpose and empowerment through embodied instinctual processes unique to the experience of motherhood, particularly breastfeeding.[45]

The Cult of Natural Motherhood and Breastfeeding

Tied to the emergence of these new theories of biological instinct and connections between mind and body, a maternal reform movement arose, centering on health practices.[46] That an early natural childbirth movement took shape in the 1940s appears almost inevitable in the context of the era's scientific discourse on the importance of mother love, biology, and maternal instinct.[47] By 1942, the emotionally detached technology of hospital childbirth that had only recently assumed prominence was already being questioned in rooming-in studies and the philosophy of Grantly Dick-Read and his followers.[48] Women became uncomfortable with the widespread practice of heavily drugged labors and forceps deliveries, and interest in alternative options grew.[49] Interest in breastfeeding gradually expanded in connection to this midcentury move toward natural motherhood practices. By the early 1950s, organizations like the Boston Association for Childbirth Education (BACE) were receiving requests from mothers and other maternal health organizations "interested in natural childbirth," and, increasingly, breastfeeding.[50] La Leche League was among the first of these groups to exclusively provide breastfeeding information and support, but it was not alone. BACE and other childbirth education organizations offered advice for mothers who wished to nurse their infants, and Catholic organizations and physicians worked to integrate the science of breastfeeding into practice. In 1958, a representative from the National Catholic Welfare Conference wrote to BACE to recommend the name of a Boston-area physician and Catholic who "boasts, much more than other obstetricians of the area, of the number of mothers he has breast feeding their children."[51] By 1962, La Leche League was just one of a dozen organizations functioning throughout the United States stressing the tenets of natural motherhood in their breastfeeding advice. An "Instructor's Kit" used by the Dallas chapter of the International Childbirth Education Association, for example, emphasized that breastfeeding was "Nature's way," and listed among the top three benefits "emotional satisfaction for her baby[,] . . . satisfaction in her own womanly and motherly functions," and "satisfaction for the father in continuing [his] role as provider and protector of his family."[52] This perspective often assumed the importance of breastfeeding for the child, while stressing the benefits for the mother, including everything from instinctual fulfillment to physical health.

Read's work was the catalyst for a broad cultural response in the United States as individual women discovered his book *Childbirth without Fear*, experienced transformative labor and deliveries using his methods, and sought to expand women's access to these options.[53] Inspired by Read, childbirth education associations began to spring up all around the country, with some of the earliest forming in Boston and the Great Lakes region. In 1960 several of these groups met to form the International Childbirth Education Association, creating a network of knowledge and resource exchange on "family centered maternity care" across the United States and Canada.[54] The group formed to join together those in "both parent and professional" communities who were interested in "preparation for childbirth, breastfeeding," and in allowing "fathers in the birth room."[55] By the mid-1960s, the Lamaze birth method, which originated in France in 1951, had overtaken the success of Read's method and was widely integrated into childbirth education classes and delivery practices. By the 1970s, historians have observed, these early groups and advocates had helped launch a widespread natural birth movement.[56]

The fact that a movement to integrate instinctual childbirth into the spectrum of choices available to mothers emerged in the postwar era is highly suggestive of a parallel history of breastfeeding, yet this connection has not been made explicit in the historiography of infant feeding. Contrary to what previous scholars have articulated, early postwar interests in natural childbirth and breastfeeding were not simply revisions of a dominant scientific motherhood ideology, nor were they a "last gasp" of Victorian motherhood. Weiner's assertion that breastfeeding offered Cold War mothers a way of imbuing their increasingly demoralized roles as mothers with a degree of dignity and authority, therefore, comes very close but misses an opportunity to identify a complex shift in American motherhood that helped give rise to an ideology of the natural. Whereas Weiner described this as a desperate and ultimately unsuccessful attempt by a handful of women in the 1950s and 1960s, I argue that not only were their efforts part of a much larger phenomenon; they were also far more successful at creating a lasting and meaningful ideological synthesis than historians have acknowledged.[57] When the knowledge and philosophies of those behind the natural birth and breastfeeding movements of the midcentury are examined more closely, it becomes necessary to conclude that an alternative maternal ideology coalesced in this era.[58] More recent investigations by scholars of motherhood and science in this period have pointed to the interwar and postwar years as a time when American motherhood experienced profound and complex changes. It is out of these broader shifts in American ideas about motherhood that a new and modern construction of breastfeeding emerged.

The Death of the Moral Mother and the Rise of the Biological "Mom"

This book engages with the expansive literature on the history of motherhood. *Back to the Breast* is not just a story about changing practices of feeding babies; it explores ways in which breastfeeding researchers and advocates maintained and expanded a body of practical and expert knowledge about human lactation, while creating an important ideological synthesis between seemingly conflicting conceptions of motherhood: moral motherhood and biological motherhood.[59] In nineteenth-century America, the sentimental love and self-sacrifice of the Victorian mother for her children defined her role as one that was both an all-consuming personal identity as well as a political one. "Mother Love," in the Victorian mind-set, had a redemptive quality and was seen as a necessity for the rearing of robust citizens, particularly sons.[60] The rise of the domestic authority of mothers in the Victorian period grew into the Progressive Era's movement of female moral reform in which the political and social efforts of women moved beyond the home to shape the public sphere on the grounds of their maternal authority and the moral superiority of their sex. Historians have attributed the decline of this moral motherhood to the growth of scientific expertise over the first half of the twentieth century, as dominion over social ills ranging from juvenile delinquency to unwed motherhood increasingly fell under the jurisdiction of professionalized experts.[61] Simultaneously, as scientific expertise usurped the role of women in the public sphere, it also followed them into the domestic sphere, as child rearing itself came to require greater scientific knowledge, technique, and assistance. Thus, the decline of the Victorian era's moral mother paralleled the rise of scientific motherhood and of bottle-feeding. In return, the ideology of scientific motherhood promised to elevate the maternal role to that of a respected technical occupation. As many works have shown, however, the rise of scientific expertise and professionalism has often worked to exclude women from positions of authority and power, even in the home.[62] Under scientific motherhood, women learned to perform their duties to the standards of scientific experts in order to be considered "good." The winnowing down of the meaning of motherhood continued over the course of the twentieth century. By the postwar era motherhood had become essentially a biological definition, a description of a particular physiological process having taken place rather than a meaningful political identity.[63]

If this extensive body of literature can tell us anything simple about the last century it is that motherhood is a construct that has been and remains viciously contested in American culture, and the mid-twentieth century was no exception. Historians have often singled out the postwar era as a period rife with sexual tension, domestic upheaval, and gender anxieties, a time characterized by female discontent and white male privilege. In this still-pervasive view of this era, the stifling repression and hegemonic ideals that stressed marriage, children, and overt consumption as all part of an "American Dream" package eventually helped provoke the widespread cultural and political backlashes of the 1960s and 1970s. The famous "problem with no name," articulated by Betty Friedan in 1962 framed an entire historical period with such resonance that her characterization of womanhood and family life in those years continues to maintain a powerful influence over the American historical imagination. The emergence of a self-conscious desire to learn about, advocate for, and practice breastfeeding among small pockets of American mothers in this period, however, stresses that Friedan's perspective did not encompass the experiences of all women, not even all of the white, middle-class housewives. Women who breastfed in this period not only challenged the status quo in the context of the health care system and their communities; they challenged the sexual and gender dynamics of their household as well. In doing so, they exerted a degree of embodied gender consciousness that is often not acknowledged by historians until the late 1960s.[64] Mothers who wanted to breastfeed received harmful advice, were ignored, and even drugged without their knowledge by a medical system built to administer bottle-feeding, not breastfeeding. However, in maintaining a focus on the small pockets of women and experts who supported breastfeeding and worked to change the outcomes for women who wanted to nurse, we ought to question the characterization of this period as one of uniform and uncritical engagement with medicine and passive acceptance of normative standards of gender, motherhood, and family.

In seeking a framework for the early breastfeeding movement, I have gratefully relied on the provocative works by two historians of motherhood and science, Rebecca Jo Plant and Marga Vicedo. Their independent analyses of motherhood and science in the mid-twentieth century offer a model of motherhood in transition in the interwar, war, and postwar years that helps better situate the early breastfeeding movement within broader historical trends. Plant has masterfully demonstrated in her 2010 book, *Mom: The Transformation of Motherhood in Modern America*, that the "mom" who emerged in the modernizing twentieth century, while more autonomous, was also a role stripped of the political, moral, and spiritual benefits afforded by the sentimentalized Victorian mother.[65] Plant captures the cultural war that erupted over the new "mom," who was expected to approach child care with a scientific rationality that demanded rigor, self-discipline, and the checking of maternal emotions that, if unleashed, might wreak havoc on the masculinity of her sons and turn her daughters into domineering domestic tyrants. It is my reading of Plant's work alongside the historiography of scientific motherhood that has led me to distinguish a particular maternal ideology that emerged in response to this change in constructions of motherhood over roughly the second third of the twentieth century. Plant herself suggests that a segment of American mothers carried on the maternalism built on the Victorian construct of moral motherhood into the post–World War II years, mobilizing cultural and political arguments from a perspective that resisted the de-sentimentalization of motherhood and yet accepted some aspects of scientific modernism. Plant posits the two conflicting views as that of moral motherhood and biological motherhood, and she suggests that by the 1960s, "mainstream American culture ceased to represent motherhood as an all-encompassing identity rooted in notions of self-sacrifice and infused with powerful social and political meaning." In place of this version of female citizenship, she argues, there emerged a view of motherhood "conceived as deeply fulfilling but fundamentally private," just one part of a female identity that by then was expected to be multifaceted.[66] One of the key components of Plant's postwar "mom," is that she became nothing more than someone who had experienced certain physiological and psychological processes. That she was by *nature* a mom, however, came to mean little about her moral character or her contributions as a citizen in general. Historian and philosopher of science Marga Vicedo has similarly described the midcentury emergence of a new kind of mother, one defined within the contexts of the sciences of maternal attachment, instinct, and bonding in the period. In Vicedo's work, the shift in constructions of motherhood is situated more explicitly within the scientific context, which she describes as a psychological redefinition of biological arguments over sexual difference. Vicedo has helped to delineate the cultural importance, furthermore, of the rise of psychological and ethological studies of "mother love" in the mid-twentieth century.[67]

Building on the theoretical contributions, in particular, of Weiner, Plant, and Vicedo, I suggest that even as biological motherhood became the dominant construction of maternal identity and experience, there remained a thread of continuity that has persisted into the twenty-first century. That story of continuity can be found most clearly in the history of breastfeeding. Breastfeeding is an act that quite literally embodies motherhood. Much like pregnancy and childbirth, breastfeeding is an undeniable marker of biological difference and provides a physiological claim for a maternal identity rooted in a secularized and material reality. Despite its intimate and personal nature, however, breastfeeding has not always been successfully limited to a "fundamentally private experience." In the resurgence of breastfeeding in the decades during and after the "death" of moral motherhood, we see the continuation of a maternal identity based on a timeless view of the moral authority of "the natural" that has persisted well into the modern era.

The Missing Link: A History of "the Natural"

In the long resurgence in breastfeeding practice over roughly the last century, an alternative ideology of motherhood emerges, an attempted synthesis of the era's biological motherhood with the same moral and cultural weight of the Victorian era. This synthesis, however, has depended on a particular understanding of "the natural" as pure, timeless, gendered female, and morally distinct from the world of male expertise, technology and authority. As such, the ideology of the natural has existed alongside but in tension with the ideology of scientific motherhood that has informed the dominant paradigm since the Progressive Era. Whereas both ideological frameworks have depended upon scientific knowledge, natural motherhood has maintained an important connection to experiential lay knowledge. Furthermore, it is reliant upon a view of motherhood as an inherently "natural" process that not only unfolds along a predetermined (evolutionarily optimized) set of biological experiences that are fundamentally similar, but also exists across boundaries of geography, culture, time, and even species. Those who helped create and spread this ideology in the period stretching from the 1920s to the 1960s notably often tended to straddle the roles of expert and mother. Whether as researchers, doctors, nurses or psychologists, maternal perspectives on science played a pivotal role in the creation and dissemination of an understanding of biological motherhood that was steeped in meaning.

The recent history of breastfeeding also brings to light a set of intriguing arguments about the connections between mind, body, and nature that mid-twentieth-century scientists and mothers relied upon to construct an ideology of natural motherhood and breastfeeding. That the argument for "the natural" should have such weight in the postwar years may at first strike readers as odd. The 1950s witnessed an unprecedented degree of confidence in scientific and medical modernity, and the number of women giving birth in a hospital under the authority of a physician was at an all-time high. Yet, there was an active and seemingly compelling discourse of "the natural" surrounding arguments about motherhood, particularly breastfeeding. To explore the meaning that "the natural" held in relation to motherhood and breastfeeding, I have turned to the work of environmental historians who have explored the history of the material connections between human bodies, nature, and the environment with great care. Conevery Valencius's 2002 book *The Health of the Country* discusses nineteenth-century human bodies and the natural world as more than just physically connected, but epistemologically and linguistically entwined as well. Valencius's work begs the question—if the bodies of those nineteenth-century Americans were so permeable and conversant with their environments, might we find threads of continuity in the experiences of the twentieth-century body as well? Environmental historians, including Chris Sellers, Gregg Mitman, Linda Nash, and Nancy Langston, have expanded these concepts into a growing and provocative historiography that, in the words of Nash, makes the early twentieth century "look like a brief period of modernist amnesia" rather than a definitive break from a less dichotomous past.[68] It is my hope that this work will bring the ideas and contributions of environmental history more directly into conversation with the history of breastfeeding. The relationship between the ideology of natural motherhood and breastfeeding in the twentieth century demands that we examine "the natural" inasmuch as we examine motherhood when we talk about breastfeeding.

Environmental historians have also been on the forefront of examining the significance of man-made chemicals as transgressors of the "nature"/"culture" border, examining them as actors in a troubling yet deeply embodied and pervasive material reality of modern landscapes. In recent years this thread of scholarship has provoked a resurgence in scholarly interest in the life and work of Rachel Carson as well as the roles played by human bodies in the awakening of a modern environmental consciousness in the period following World War II. Researchers learned soon after the widespread implementation of the pesticide DDT, for example, that breast milk acted as a biological magnifier for chemical toxins in the body.[69] Despite the connections between polluted environments and polluted bodies that became increasingly known in the postwar years, many breastfeeding advocates remained committed to a concept of breastfeeding as natural. That breastfeeding's fundamental worth seems to be tied to its existence as a *natural* process adds further strength to its connection to an underlying ideology rooted in ideas of nature as "pure" and yet no longer separate from ourselves.

The 1940s and 1950s were not the first decades to rely on a concept of breastfeeding as "natural." Arguments for breastfeeding as natural, in fact, can be traced far back into our history. Londa Schiebinger's account of the eighteenth-century biologist Linnaeus and his classification and naming of mammals (after "mamma," the glands responsible for lactation), for example, demonstrates the extent to which the feeding of one's young via the breast has long sat at the core of how Western civilization has constructed motherhood and breastfeeding as fundamentally "natural," so much so that it could serve as a category for scientifically organizing and naming the world's creatures.[70] What I suggest, rather, is that breastfeeding became something that could reforge a deeply yearned-for connection with nature. In turning back to breastfeeding, some mothers looked to their biological ability to nurse as a source of meaning and solace, an embodied refuge.[71] The physical and emotional act of breastfeeding reemerged in the second half of the twentieth century as a circumscribed, imperfect, and embodied form of nature, not unlike the nature the largely white, middle class cultivated in its own suburban backyards.[72] Breastfeeding mothers in this era were some of the earliest consumers of scientific information about environmental toxins and some of the most sophisticated. Cultivating this link to a nature both timeless and troubled became something of increasing value to women, particularly in the years after World War II.

In this period some women began to breastfeed because it was a way for them to embody a particular concept of nature that they valued, a perspective that included cultural and moral arguments for the utility and value of scientifically described behavioral and physiological instincts felt in and by the maternal body. The ideological framework that reinvigorated breastfeeding with new meaning in the postwar years was not simply about ethological theories of instinct, cultural longings for a Victorian golden age, or maternal resistance in an age of limiting freedoms for women. The ideology of natural motherhood grew out of an evolving philosophical understanding of the relationship between humans and the natural world, a nature understood to be deeply embedded in the bodies of women.

A Note on Race in the History of Breastfeeding

The majority of the narrative in this book deals with the experiences, resources, and worldview of a small segment of America's white middle-class mothers. The researchers who studied maternal behavior and infant development in the mid-twentieth century were overwhelmingly united in their whiteness. The women who shaped the early movement back to the breast were able to do so largely because their racial and economic privileges provided them with access to educational resources, community support, and medical competence. Not unlike their maternalist forebears of the early twentieth century, the mothers who influenced breastfeeding knowledge, support, and policies in the second half the century carried their middle-class assumptions and ideals with them wherever they went. Throughout the first several decades' of La Leche League's existence, for example, the leadership consistently articulated that working while mothering was a "choice," and it was not, in their opinion, the best choice. Black mothers, however, had their own complicated relationship with breastfeeding that went far beyond the issue of working. The long history of white families relying upon black women's labor as wet nurses and child nurses into the twentieth century, for example, contributed to a lingering disconnect between black motherhood and breastfeeding that continues to percolate in contemporary discussions by and about black mothers.[73] That black mothers in the United States have breastfed at historically lower rates since the resurgence in the practice began in the 1970s makes perfect sense given that for many black women breastfeeding remained embedded within a history and ongoing experience of racial oppression.[74] Just as salient to the history of the relationship between race and breastfeeding is the long tradition of a racist scientific and medical discourse, which helped lend legitimacy to constructions of blacks, for example, as genetically inferior, culturally primitive, and evolutionarily closer to animals than whites.[75] For black mothers, therefore, connections to the natural could often be tinged with meanings of racial inferiority and gendered exploitation, rather than empowerment. For white Americans, an interest in connecting with nature had long come at the cost of the implementation and maintenance of a racial hierarchy. At the turn of the twentieth century, concerns over the strength of white masculinity fueled arguments that white men and boys needed to exercise aspects of their "inner savage" through hunting and other outdoor conquests. White mothers at midcentury similarly faced a crisis in maternal authority and responded by embracing pieces of "the natural" and "primitive" for themselves. Cultural and racial "others" served as subjects in anthropological reports and scientific studies throughout the mid-twentieth century, which in turn served as sources of information and inspiration to white middle-class mothers looking to live more "naturally."[76] For black American mothers, however, long-standing associations between breastfeeding, exploitation, and the primitive often made ideologies of the natural unpalatable and culturally irrelevant.[77]

Yet despite all this, by the 1980s breastfeeding rates among nonwhite American mothers were also on the rise. Although today there remains a considerable gap in the rates of breastfeeding between white and black mothers in the United States, all segments of American society participated in the move back to the breast by the century's end.[78] While this book does not detail the unique experiences of black mothers or other minority mothers (Hispanic, Asian American, lesbian, etc.), it does suggest that by the turn of the twenty-first century, there were very few mothers in America who would have felt untouched in one way or another by the back-to-the-breast movement and the ideology of natural motherhood.

Perspective

It is worth stating outright that this is not a book about formula feeding, bottle-feeding, or infant feeding more generally. Though I have often been asked whether I am pro-breastfeeding or pro-bottle-feeding, this book is not my attempt at weighing in on the "breast versus bottle" debate nor is it my desire to make moral or scientific claims about the superiority or neutrality of breastfeeding or formula feeding now or in the past. Instead, this is a book that explores the story of how, why, and to what effect a particular ideological construction of breastfeeding and motherhood as natural became meaningful, important, and popular after a long period of decline. My work on this project has informed my position that every mother should have the ability to make a real choice about how to feed her baby, one that is as free from constraint as possible (economic, familial, physical, social, or otherwise) and supported equally. Throughout much of the last century, limitations on women's abilities to exercise choice regarding breastfeeding have been so extreme as to negate the viability or utility of the rhetoric of "choice" for many women. This was particularly so throughout the 1940s–1970s, when societal and institutional obstacles to breastfeeding were so widespread that women had to form an advocacy and support organization just to let mothers know that they were capable of breastfeeding. In this context, the mother who breastfed her baby for more than a few weeks was an anomaly and in many ways a radical. I do not mean to discount the experiences of those women from throughout this same period who have also exerted agency and experienced a sense of empowerment through early weaning or formula feeding their infants. Similarly, it is certainly not my contention that natural motherhood has been the only ideology shaping American motherhood or breastfeeding throughout the past century, nor was it necessarily the most important or dominant one to all women at all times. What I do articulate throughout *Back to the Breast* is that the ideology of natural motherhood shaped the path of breastfeeding's return to popularity in every way. The marriage of breastfeeding to the ideology of natural motherhood was an important component in the early back-to-the-breast movement and this fact has continued to have meaningful implications for breastfeeding practice up through today. While it has competed with other ideological constructions in shaping ideas and practices surrounding breastfeeding over much of the last century, natural motherhood has fundamentally influenced how Americans today have come to think about breastfeeding. In the chapters that follow, I focus on the persistence in the belief by countless Americans over the past century that breastfeeding holds value and meaning that transcends nutritional adequacy and infant survival. I trace the efforts, science, struggles, triumphs, and failures of the people and ideas behind the back-to-the-breast movement over much of the last century so that we might better imagine a society in which all mothers receive the support they need to make their experiences as mothers personally rewarding and fulfilling.

CHAPTER ONE

Make Room for Mother

The "Psy"-entific Ideology of Natural Motherhood

In 1928, psychologist John B. Watson launched what would become one of the best-selling child care advice manuals of the twentieth century, inscribed with its now well-known dedication to "the first mother who brings up a happy child."[1] Watson's sarcasm reflected the era's vitriolic skepticism of nonprofessional female expertise. By the late 1920s mothers who resisted the march of scientific progress by following the advice of grandmothers and "old wives' tales" were at best mocked as ignorant and at worst considered threats to the very survival of their children, and even civilization. Less than two decades later, a different kind of advice manual was slowly making its way into the hands of American mothers. In 1944, *Childbirth without Fear: The Principles and Practice of Natural Childbirth*, began attracting a small but avid following in the United States. The British author and obstetrician Grantly Dick-Read assured his readers that the female body was inherently capable of birthing a baby and that the pain, fear, and danger that accompanied modern childbirth resulted from the psychological conditioning society imposed on women. The difference in their messages is stark: in Watson's eyes mothers were clueless dangers to their children without the competent guidance of a scientific expert, while Read's philosophy suggested that women were *naturally* mothers. By the 1940s, Read was just one of a growing international network of scientific experts who believed that the main obstacle women faced in becoming successful mothers was the intervention of modern scientific society itself.[2]

This chapter explores how the midcentury human and biological sciences helped construct a modern ideology of natural motherhood through the study of mother-infant interactions across boundaries of difference, including human/animal distinctions, disciplinary and national borders, and cultures. An international body of scientific and clinical work laid the foundation for natural motherhood between the 1920s and 1950s. While the ideology of scientific motherhood, which had helped fuel the rise of scientific formula and bottle-feeding, remained a dominant force in American motherhood throughout this period, it also helped provoke a backlash in science and culture that stressed the importance of emotion and maternal love in child rearing.[3] By the 1940s the earliest of these works was already shaping a popular movement in support of natural motherhood in both childbirth and breastfeeding practices. Throughout my discussion of these scientists, I draw on the apt terminology of Nikolas Rose whose concept of the "psy"-ences captures the indefinite boundaries that continued to characterize the mid-twentieth-century fields of psychology, psychiatry, anthropology, and even ethology, the science of animal behavior.[4] These experts contributed to the construction of a uniquely modern and scientific ideology of natural motherhood. In doing so, they helped redefine the meaning of breastfeeding for mothers in the second half of the twentieth century.

The Psy-ence of Natural Motherhood

At its most basic, the ideology of natural motherhood refers to a body of scientific models, theories, and teleological assumptions as well as widespread cultural beliefs that natural instincts of both an emotional and a physical nature govern maternal behavior and infant development. As complex as this definition can become when all of its various components are explored, its distillation yields a simple and very real point of difference between the competing ideologies of natural and scientific motherhood: natural motherhood relies upon a model of *embodied maternal knowledge* while scientific motherhood relies on the external knowledge and technologies of scientific experts. How a mother accesses this knowledge is a question that has evoked different answers depending on time and circumstance, but the concept of an embodied source of wisdom in infant rearing offered a dramatic departure from the model of knowledge transmission in scientific motherhood.

The pages that follow explore the shift in the psy-ences toward a central concern with the psychological and emotional influence of the mother on her child. Beginning with Watson in the United States and extending to the legacy of Sigmund Freud through the work of his daughter, Anna Freud, and her disciplinary rival, Melanie Klein, I explore how psychologists and psychoanalysts thought and wrote about mother-infant relationships, paying particular attention to breastfeeding. I then trace the impact of the work of the British psychoanalyst, John Bowlby, whose interest in breaking out of the early twentieth-century psychoanalytic debate over infant feeding drove him to explore the work of ethologists and experimental psychologists. While Bowlby and others in his camp, including psychologist Harry Harlow, worked to disconnect infant psychology from the subject of feeding by identifying a more powerful and ultimately more primal internal drive than physical hunger, the work of these men in articulating "love" as an instinctual physical and psychological need ironically provided the foundation for a scientific argument in favor of breastfeeding. In the final section of this chapter, I discuss the lives and work of two highly influential female psy-entists who helped translate this work through both their own lives as women and mothers, and as experts. Throughout the World War II and postwar eras, anthropologist Margaret Mead and her student, psychologist Niles Newton, helped shape the work of male psy-entists like Bowlby into a maternal-centric ideology of natural motherhood in which breastfeeding symbolized a powerful and autonomous female identity based on embodied natural instincts and a belief in universal womanhood.

The Psychology of Mothering

John B. Watson's work helped fundamentally shape the role of American psychology in the study of infant rearing and motherhood. His 1928 best-selling child care advice manual *Psychological Care of Infant and Child* integrated his theories of Behaviorism seamlessly into the day-to-day care of a child, stating simply that "since children are made not born, failure to bring up a happy child, a well adjusted child—assuming bodily health—falls upon the parents' shoulders."[5] Watson's assertion that "no one today knows enough to raise a child" spoke directly to the early twentieth century's scientific motherhood. This powerful ideology held women responsible for child rearing but also required that they seek "expert advice in order to perform their duties successfully."[6]

Watson's work, perhaps ironically, foreshadowed a change in perspective in mother-infant study that would challenge the hegemony of scientific motherhood. In *Psychological Care*, Watson articulated the ideal of the professional mother, a model of womanhood in which mothers deferred to male scientific authority on domestic matters but also devoted themselves fully to the job of rearing children, scientifically. Men like Watson wrote passionately about the shortcomings of American mothers with the hopes of guiding women to accept the superiority of scientific knowledge in the maternal sphere and to eschew the coddling of the Victorian matriarch. With well-practiced condescension, Watson berated American mothers for their ignorance and denounced the role of "nature" in mothering. For most mothers, he wrote, "the age-old belief that all that children need is food as often as they call for it, warm clothes and a roof over their heads at night, is enough." "'Nature,'" he wrote mockingly, "does the rest almost unaided."[7] In Watson's eyes, and in the eyes of many of his peers, "nature" was not a guide but an enemy, an ideological holdover from the sentimentalist arguments of the Victorian era when most believed women were inherently gifted with the ability to raise children. This view of motherhood notably highlighted the lack of psychological awareness that *un*scientific mothers possessed. Such backward women, Watson mocked, continued to believe that feeding was the most important part of raising an infant while they neglected their children's psyche and emotional health.

While Watson argued for the expansion of scientific motherhood to include psychology among the more established areas of hygiene, nutrition, and regimen, he also set the science of motherhood on a new path. By the 1920s, Watson's work had helped popularize the idea that in addition to overseeing basic physical care, mothers needed to raise mentally and emotionally fit children in order to adequately fulfill their roles. While historians have highlighted the extent to which this contributed to the creation of unrealistic and harmful expectations of modern mothers, Watson's acknowledgment of the importance of emotions in child rearing had other unintended consequences as well.[8] Despite the behaviorists' disbelief in human instinct and the idea of "nature," the idea that feelings were an important part of child rearing necessarily introduced the subjective and the experiential into the developing science of motherhood. This would go on to create the possibility for a new brand of maternal expertise altogether, based on a new kind of psy-ence that took "female" ways of knowing seriously again. Even as mothers came under increasing surveillance by the psy-ences as the century wore on, therefore, they also gained access to a set of arguments and explanations that supported a new model of maternal authority.

As psy-entists scrutinized the emotional fitness of mothers, their gaze also appropriately shifted to the psychology of infancy. In 1938, Dr. C. Anderson Aldrich, an associate professor at Northwestern University Medical School and pediatric clinician (and later, the mentor of Dr. Benjamin Spock), publicized a new way of thinking about infancy. His influential book, *Babies Are Human Beings: An Interpretation of Growth*, laid out in layman's terms an evolutionary model of infant growth and development that integrated biological theories of childhood growth with mental development. His rhetoric throughout emphasized that "nature," in fact, played an important role in the raising of an infant. Aldrich's work drew direct links between infant feeding and a concept of "natural" development in human babies. He suggested that infants innately know to cry out when hungry; therefore, it made sense that "in order to help this pain-food-relief sequence which makes eating pleasant, we supply food when we hear the hunger cry." He also noted that the infants' pleasure in eating could only be enhanced by ensuring that "the surroundings of his early meals are made as comfortable as possible." Luckily for humans, he added, "Nature provides an ideal set-up for this when she arranges that he is cuddled and warmed by his mother while nursing."[9] Unlike the earlier claims of the behaviorists, Aldrich helped articulate an argument for a biological-psychological system of feedback that linked infant and maternal behavior into a harmonious rhythm, choreographed by nature. In Aldrich's view, nature and nurture worked in concert, giving purpose and meaning to maternal-infant interactions, particularly feeding.

The work of Dr. Margaret Ribble, who spent four years studying in Vienna with Anna Freud, also contributed to a growing canon that emphasized early mother-infant interactions as foundational for psychological and emotional development in both the infant and the mother. Ribble's work stressed the importance of infancy as a period of critical psychological development that depended almost exclusively on "woman's instinctual nature."[10] In her 1943 popular crossover work, *The Rights of Infants*, Ribble critiqued mainstream psychological perspectives on mother-infant dynamics in favor of one that embraced the emotional and instinctual aspects of mothering more directly.[11] She wrote, "Modern science, when it considers the matter, assumes that this basic tie [between mother and infant] exists in order that the child may be fed and protected from harm during its helpless infancy." For Ribble, this theoretical platform diminished the role of the mother, turning her into nothing more than "a trustworthy nurse, who can arbitrarily be replaced." She called for a more complete consideration of "the matter of a personal relationship" between mother and infant, one "on which the child's future emotional and social reactions are based.[12]

Ribble's work also assured women that they could learn "both by instinct and observation" how to care for their babies and argued "that ['mothering'] is as vital to a child's development as food."[13] In her lengthy discussion of breastfeeding, Ribble declared that it was "of the very essence of 'mothering' and the most important means of immunizing a baby against anxiety." She emphasized, furthermore, that when it came to infant feeding it was best to "follow Nature's cue," because in "normal breast feeding" (which, she added, should continue "at least until the teething period"), "the various instinctual hungers are self regulated."[14] These "instinctual hungers" were best met, she articulated again, and again, through the mother's natural ability to breastfeed. She went on to emphasize: "Breast-fed babies tend to have more trust and confidence in their mothers."[15]

In her writings, Ribble not only focused on the importance of a successful breastfeeding relationship; she also highlighted the period immediately following birth as a crucial time in the infant's psychological development. A 1945 *Parents' Magazine* article about Ribble's work explained to its readers that "breast feeding is . . . an important part of mothering, giving as it does the first opportunity for cultivating the child's emotional capacities." "Every time you cuddle your baby in your arms and release your milk for his nourishment," the author added, "you add to his sense of well-being and security."[16] In the early 1940s, Ribble sat at the forefront of a psy-entific paradigm that would increasingly emphasize the crucial importance of the physical and emotional circumstances of the immediate postpartum period for long-term psychological development.

The works of Ribble and Aldrich, which identified the existence of a complex mental life in the infant from birth, represented views that became increasingly common among psy-entists throughout the 1930s and into the early 1940s. That psy-entists considered the mental state of the infant at all was a divergence from an earlier school of thought notably championed by Rousseau, which held that little mental development occurred prior to the "age of reason," around seven or eight years old.[17] Sigmund Freud's work on Oedipal experiences and sexual pleasure drives played an integral part in this eventual shift to a concern about the early inner life of infants. For example, a widely cited article published by analyst David Levy in a 1928 issue of the *American Journal of Psychiatry* explored the contentious issue of finger sucking through a Freudian framework, suggesting ultimately that it occurred as the result of problematic "feeding methods."[18] In a series of twenty case study investigations, Levy explored finger sucking and "masturbation" in toddler-aged children, tracing each of their so-called pathologies to the method of their early feedings at their mothers' breasts. In one case he observed that an exclusively breastfed baby girl was, on the first feeding, "pulled away from the breast and was [thereafter] never allowed to nurse 15 minutes."[19] According to Levy, this resulted in an unsatisfied pleasure drive, causing the child to pathologically suck her thumb as a toddler.

The work of Freud's daughter, Anna Freud, helped to further the shift in psychoanalytic interest to the early period of infancy. Anna Freud helped contribute to the understanding of a developmental arc in the mental life of an infant that was entirely dependent upon a close relationship with his or her mother. Her work in wartime nurseries during World War II showed that even brief periods of separation between infants and their mothers at key developmental stages could have long-lasting mental and physical effects on child development. In her exploration of infant and child psychology, Freud turned to infant feeding behaviors as a tool of inquiry. The results of her 1946 Feeding Behavior Study further bolstered the psychoanalytic view of the importance of feeding experiences for long-term infant development.[20] Anna Freud's work helped support the relevance of her father's theories for feeding and infant development. She argued that it was the pleasure of the feeding experience itself that the infant invested with love "and only later with the growth of awareness does the infant cathect [love] the milk, breast or bottle . . . and finally the mothering person."[21] Freud's analysis of this process, however, only added greater tension to the question of breast versus bottle because her work stressed the importance of early infant feeding as the root of maternal-infant love and psychological growth.

Another unique, yet parallel, school of thought emerged out of Freudian psychoanalysis in the 1930s in the work of Melanie Klein. Herself the mother of three children, Klein entered the world of psychoanalysis with an explicit interest in child analysis. Still a relatively new field when she entered it in the 1910s, the psychoanalytical study of children became increasingly popular by the postwar years. By then, Klein had successfully established herself thanks to the success of her work in object relations theory. Object relations theorists argued that babies very early in life began ordering their environment into "good" objects and "bad" objects in order to deal with the emotional and physical frustrations of infancy. The primary object in Klein's early models was the mother's breast. As Klein wrote, "In tracing the analyses of adults and children, the development of impulses, phantasies [*sic*] and anxieties back to their origin, i.e. to the feelings towards the mother's breast (even with children who have not been breast-fed), I found that object relations start almost at birth and arise with the first feeding experience."[22] The Kleinian perspective also emphasized feeding while it simultaneously created the possibility of neutrality on the bottle/breast controversy.

Both Freud and Klein maintained a focus on the maternal breast, or its substitute the bottle, as the primary motivator in infancy. The only way in which the two notably differed on this point was in their understanding of the actual process through which an infant became attached to its mother. Freud insisted that the physical pleasure the infant experienced while breastfeeding was the key to psychological development over time. Klein, on the other hand, ascribed to infants a more complex internal psyche from birth, in which the infant was capable of emotional and psychical motivations in addition to physical and libidinal drives. Object relationists, therefore, saw the breast or the bottle itself as a fundamental and symbolic part of the infant's psychological mapping of its environment.[23] While controversial and meaningful for the British psychoanalytic community at the time, these differences between the two women had little differential impact on the meaning of breastfeeding in general. Followers of both Anna Freud and Klein agreed that feeding was an integral part of infant psychological development.[24] Additionally, both theorists allowed room for the bottle to serve as a reasonable substitute for the breast. Notably, Klein's earlier works from the 1920s and 1930s took breastfeeding for granted as the most common means of feeding, but by the 1950s her phrasing changed to discussion of breast or bottle-feeding as interchangeable.[25] Because the infant's attachment (Freud) or psychological mapping (Klein) would eventually transfer from the immediate source of its nourishment to the whole mother, the method in which an infant was fed mattered less than the *manner* in which it was fed. This allowed for the integration of bottle-feeding into dominant psychoanalytic thought in a way that did not necessarily leave the bottle-feeding or breastfeeding mother automatically exposed to negative or positive judgment. In abstaining from any pronouncements on the means of infant feeding, Freud and Klein walked in what was by then a well-worn path. Psychoanalytic opinions on feeding method had long been ambiguous if not completely neutral. In Levy's 1928 study on finger sucking, for example, he had observed that the ultimate cause of the undesirable behavior in toddlers was some sort of abrupt change in the feeding experience, whether it was due to a forced withdrawal of the breast or the bottle or a shift in feeding schedule that created anxiety in the infant. "There is no relation of fingersucking to the type of feeding," he observed, "breast, bottle, or mixed." He believed it made little difference in the infant's psychological development.[26]

Though she had studied with Anna Freud, the work of Margaret Ribble offered some of the earliest challenges to this neutral view, by proposing that the importance of maternal nurture to the infant's early psychological development actually favored breastfeeding. A graduate of Cornell's medical college, Ribble utilized psychological and biological arguments to construct a model of the mother-infant pair in which physical processes had lasting psychological consequences, and vice versa. In one of her earliest works she argued that an infant's emotional and psychological "appetite" for sucking at the breast preceded its actual nutritional hunger.[27] The sucking of the infant at the breast, likewise, elicited certain responses in the mother, including talking and stroking of the infant's face. This process unfolded over a series of weeks. By the end of a two-month period, she observed, "the infant looks at the mother at the sound of her approach," marking a "definite recognition" in the infant's mind of his or her mother.[28]

Despite the traction of Ribble's ideas among popular audiences at the time, she confronted a series of professional attacks by American psyentists who took issue with her work and excoriated her in their journals.[29] Her attempts at marrying scientific authority with maternal authority often provoked ire from antisentimentalists who overlooked the scientific nuance of her work and instead viewed her ideas about "mothering" as old-fashioned, a critique that followed her to her death.[30] When she died in 1971 her obituary appeared in the *New York Times* and in her local paper in Warrenton, Virginia, the *Free Lance-Star*, which memorialized her briefly as a "New York City psychiatrist" who had been a "member of Phi Beta Kappa and studied four years in Vienna on fellowship."[31] Her own correspondence with her mentor, Anna Freud, however, reveals that throughout her career she juggled a busy public speaking and publication schedule.[32] Her letters to Freud also suggest that she struggled with a troubled emotional life that often hinged on her angst over her identity as a mother and her troubled relationships with her children and grandchildren.[33] Despite her struggles later in her career, Ribble's work struck an important chord with many women in the 1950s. While her professional peers critiqued her for her perspectives on motherhood, her work received attention from popular audiences, particularly mothers. That this was the case suggests not only that female psy-entists helped to create alternative constructions of modern motherhood and breastfeeding, but that American mothers were often eager to embrace these alternatives as they sought meaning and satisfaction through their biological roles as mothers.

Infants, Instincts, and the Evolution of Attachment Theory

World War II offered psy-entists new opportunities and resources for studying the relationship between motherhood and infant development. Anna Freud and Dorothy Burlingham, for example, were among the first to explore the impact of war and institutional care on children. At the beginning of the war, they opened the Hampstead nursery to house and study orphans and infants whose mothers were employed in the war effort. René Spitz took the study of institutionalized infants further when he filmed orphaned and neglected infants and toddlers in the nursery of a women's prison outside New York City and in a foundling home in Central America.[34] The result of his efforts received widespread attention, resulting in a rising call for institutional reform in America, but Spitz's work also carried significance for child rearing more generally. Spitz's film contributed to a growing uproar over the consequences of institutionalized childhoods and the effects of inadequate mothering. In his film *Psychogenic Disease in Infancy* (1952), Spitz depicted emotionally neglected infants and toddlers, confined to cribs with only cursory contact with a rotating staff throughout their days. They had little stimulation in the way of toys or access to other people for any length of time, and were pictured feeding themselves with bottles propped on pillows or blankets for those too young or weak to hold them themselves. The infants were shown over the course of a year, the compressed timeline highlighting their decline into regressive behaviors and a withdrawal into themselves from the world around them. Fecal smearing and the ingestion of their own waste were some of the more disturbing behaviors exhibited, while most of Spitz's subjects seemed to eventually succumb to an almost vegetative state, a condition he referred to as "anaclitic depression" and "hospitalism."[35] He described the condition as a devastating and irreversible developmental catastrophe in which the child not only stopped developing along the normal arc, but actually regressed, showing reduced physical and cognitive capacity and an inability to form appropriate social attachments, damage that would last throughout their lives. The longer a child suffered the absence of a mother figure, argued Spitz, the more severe and less reversible their psychological condition.[36]

Alongside Freud and Burlingham's research, the work of Spitz was frequently called upon as an argument for better and more intensive mothering for all infants. As one author wrote in 1956, "The work of Spitz, Burlingame [*sic*], and Anna Freud has made clear that the normal healthy growth of a child requires a close, warm, affectionate relationship with the mother."[37] In the same vein, however, these experts' work was just as often utilized in discussions of neutrality when it came to breastfeeding. The same author of the above passage, for example, also commented:

> One hears almost endless heated argument about breast feeding. There are those who would have us believe that it is nearly as heinous to bottle feed as it is not to feed at all! We need to have some good common sense about this. . . . The really important thing is clearly how the mother *feels* toward the baby. If she really loves the baby and welcomes her motherhood, it is then a matter of what makes *her* feel most comfortable with her baby.[38]

Despite their interest in infant feeding as a subject of study, psychoanalysts generally remained neutral when it came to pronouncements on what to feed babies. The theories of Freud and Klein, after all, emphasized feeding problems, but with the exception of Margaret Ribble, neither school focused much attention on the differences between bottle and breast. This would change as psy-entists began to utilize studies of animal behavior in the growing field of ethology to construct a model of motherhood based on natural instinct.

In the 1930s, before psychologists ever had reason to carry out their wartime investigations into the effects of loss and maternal separation on children, a burgeoning field in biology was taking shape in Eastern Europe. The work of Konrad Lorenz (in Austria and Germany) and Nikolaas Tinbergen (in the Netherlands) was beginning in the world of ethology, the study of animal behavior. Their work on the behavioral development of birds led them to their Nobel Prize–winning theories of social behavior and organization in animals.[39] Despite their disciplinary distance, psychologist John Bowlby, described by a biographer as a "keen naturalist," was incredibly intrigued by the 1952 translation of Konrad Lorenz's work on attachment in *King Solomon's Ring*. Bowlby's notes to himself on his reading of Lorenz's work illustrate his interest in the process of maternal-infant bonding and its relationship (or non-relationship) to feeding: "Newly hatched goslings follow their mother (or a mother-surrogate), and exhibit analogues of 'anxiety' (cheeping, searching) when separated from her, despite the fact that she does not directly provide them with food. Here bonding seems to be dissociated from feeding."[40]

Looking for evidence that would support his belief that something other than food had to be at the root of infant behavior, Bowlby eagerly made note of ethological findings like these that not only offered scientific validity to his own ideas about the human experience, but also drew the investigation of infant development into the world of scientific research and evolutionary biology. In a presentation on their work, Dr. Mary Ainsworth (his student and colleague) discussed the impact of Konrad Lorenz's work, and ethology in general, on Bowlby's thinking: "He found the descriptions of separation distress and proximity seeking of precocial birds, which had become imprinted on the mother, strikingly similar to those of young children. He was also struck by the evidence that a strong social bond can be formed that is not based on oral gratification."[41]

In ethology, Bowlby could distance himself from what he saw as the limits of the developmental theories of his psychoanalytic peers. As he and Ainsworth noted, "Bowlby disagreed with the psychoanalytic theorists who, like Melanie Klein, believed that the loss of the breast at weaning is the greatest loss in infancy."[42] Furthermore, Bowlby's discovery of the field of ethology in his own work led him to delve more deeply into the literatures coming out of evolutionary biology and systems theory. The perspectives he found there greatly influenced his growing belief throughout the 1950s in the parallels between animal and human behavior. He maintained this perspective throughout his career, writing in 1969: "Whatever behavior is found in sub-human primates we can be confident is likely to be truly homologous with what obtains in man."[43]

As Bowlby biographer Jeremy Holmes has written, in the spirit of professionalization via scientization, John Bowlby sought to bring twentieth-century scientific practice to bear on the field of psychoanalysis. Disinterested in observing the disciplinary boundaries that his mentors in the object relations school advocated, Bowlby looked to biology for ways to provide scientific validity to the psychoanalytic theories he practiced. Schooled under Klein, Bowlby found that his own experiences as a child who was sent away to boarding school at the age of six, along with his professional work with juvenile criminals, did not exactly overlap with what the Freudians or the Kleinians articulated. In 1950, Bowlby's work in the area of maternal separation and mental health earned him a job with the World Health Organization (WHO), which commissioned him to report on the status of children in Europe and the United States after the war.[44] The result was an immensely influential survey of the literature on the impact of separating young children from their mothers, titled *Maternal Care and Mental Health*. In it, Bowlby summarized the state of scientific knowledge regarding the mental health and development of children. He wrote, "For the moment it is sufficient to say that what is believed to be essential for mental health is that the infant and young child should experience a warm, intimate, and continuous relationship with his mother (or permanent mother-substitute) in which both find satisfaction and enjoyment."[45] Throughout his report, Bowlby emphasized the importance of a consistent, attentive, and affectionate mother figure as an essential part of normal child development. He was specific in highlighting that this primary caregiver should be a woman. While he allowed that fathers had their role, Bowlby carefully emphasized that "it is this complex, rich, and rewarding relationship with the mother in the early years, varied in countless ways by relations with the father and with siblings, that child psychiatrists and many others now believe to underlie the development of character and of mental health."[46]

Bowlby helped unite ethological, psychiatric, and psychological thinking on child rearing and infant care by synthesizing and packaging his theories in such a way as to make them appear almost obvious. For example, in the excerpts quoted above, Bowlby easily turned assumptions of the importance of early childhood development (i.e., infancy) into a statement of scientific fact. He wrote, "The quality of parental care which a child receives in his earliest years is of vital importance for his future mental health."[47] Perhaps more so than his earlier scientific discussions of infant attachment, Bowlby's discussion of "maternal deprivation" in the WHO report resonated widely with the American public. He wrote: "A state of affairs in which the child does not have this relationship is termed 'maternal deprivation.' This is a general term covering a number of different situations. Thus, a child is deprived even though living at home if his mother (or permanent mother-substitute) is unable to give him the loving care small children need."[48]

Bowlby's definition of "maternal deprivation" differed dramatically from earlier versions of emotional neglect, most notably the "hospitalism" and "anaclitic depression" articulated in Spitz's work. These earlier conceptions of deprivation in infancy had centered on the failures of institutions, particularly orphanages and nurseries, which left infants in almost complete isolation from meaningful human contact. Bowlby expanded the term from its institutional origins to include the idea of *insufficient* mothering. In doing so, he provided a scientific basis for a higher level of scrutiny of average American mothers and their relationships with their children. "When deprived of maternal care," he warned, "the child's development is almost always retarded—physically, intellectually, and socially—and . . . symptoms of physical and mental illness may appear."[49]

Bowlby's WHO report had an immediate impact both in science and among the public. Popular audiences consumed his work directly after the WHO report was republished with the title *Child Care and the Growth of Love* in 1953. The popular version sold 450,000 copies in English and was translated into ten languages.[50] A 1963 retrospective in the *New York Times* characterized the response to the report as follows:

> The original Bowlby report, it seems, has influenced a great many parents who heard and believed implicitly that it is essential for a child "to experience a warm, intimate and continuous relationship with his mother (or permanent mother-substitute)" from birth. Many fear that if this relationship does not exist, the child will suffer from "maternal deprivation." And unless completely insulated from current thinking on this subject, they have come

> to believe that "maternal deprivation" has grave, far-reaching and permanent effects on the young, leading to social backwardness in infancy, juvenile delinquency in adolescence and all sorts of neurotic behavior in young adulthood. . . . "Maternal deprivation," however, soon became applied to *all* situations in which mothers were separated from their children.[51]

As this particular author lamented, follow-up studies into the 1960s showed that the effect of Bowlby's research had been to drive mothers into the realm of "smothering." Citing the follow-up publication by the World Health Organization, the author suggested "many youngsters have been so overwhelmed by ever-present, doting mothers that they remained infants even at the age of 6 or 7."[52] As the 1963 article suggested, however, maternal deprivation resonated with Americans throughout the 1950s and into the 1960s, a phenomenon that suggests a persistent and widespread concern over mother love.

In a 1958 article published in the *International Journal of Psycho-Analysis*, Bowlby provided a sweeping and masterful review of the literature up to that point on "the nature of the child's tie to his mother." Drawing from the traditional psychoanalytical literature of Freud, Klein, Ribble, and others, in addition to citing and discussing Charles Darwin's work on instinct and Harry Harlow's work on rhesus monkeys, Bowlby articulated both the need and the rationale behind his universal hypothesis for child development. Bowlby argued that there were five instinctual and uniquely human infant responses that unfolded in a developmental pattern in such a way as to focus the infant's affections and psyche ultimately on the specific form of his own, unique mother. "It will be my thesis," he wrote "that the five responses which I have suggested go to make up attachment behavior—sucking, clinging, following, crying, and smiling—are behavior patterns of this kind and specific to Man."[53] In citing a series of instinctual behaviors leading up to the creation of unique and person-specific attachment to the infant's mother, Bowlby challenged a multitude of theories and conventions in psychoanalysis that had emphasized feeding in infant study up until that point. First, in utilizing biological and ethological research to construct his theory, Bowlby challenged the bounds of traditional psychoanalytic theories, and second, in his elucidation of a series of innate physiological responses in the infant, he attempted to decouple the study of infant development from infant feeding. While Bowlby included "sucking" in his list of instinctual developmental behaviors, he was adamant about placing it in its proper context, as just one of several important "mother-oriented instinctual responses." "I have left *sucking* to the last," he wrote, explaining, "My reason is that psycho-analytical theory has tended to become fixated on orality and it is a main purpose of this paper to free it for broader development."[54] Ultimately, he argued, it was the love and affection that an infant received from its mother that mattered, not breast or bottle. He wrote, "It is my thesis that, as in the young of other species, there matures in the early months of life of the human infant a complex and nicely balanced equipment of instinctual responses, the function of which is to ensure that he obtains parental care sufficient for his survival."[55] Perhaps even more importantly, Bowlby articulated this argument within the biological framework of instinctual response. The idea that a mother and infant existed within a feedback loop with both physiological and emotional responses contributed greatly to the idea of natural motherhood. Citing the work of Konrad Lorenz, Bowlby argued, "as human beings, we experience the activation in ourselves of an instinctual response system." "When the system is active and free to reach termination," he added, "we experience an urge to action accompanied . . . by an emotional state peculiar to each response."[56] Expounding on this, Bowlby challenged the interventionist ethos of modern scientific motherhood when he suggested that these instinctual response systems could be interrupted, leading to ill effects. "When . . . the [instinctual] response is not free to reach termination, our experience may be very different: we experience tension, unease and anxiety."[57] Despite Bowlby's argument against the importance of infant feeding, his framework of maternal-infant instinctual response actually provided a much stronger argument in favor of breastfeeding. If physiological and emotional responses in the infant triggered physiological and emotional responses in the mother in a continuous loop of psychobiological dependency, then interfering in biological maternal-infant processes (i.e., childbirth and breastfeeding) could have negative psychological consequences for both the infant and the mother.

In researching his article, Bowlby came across the work of the American psychologist Harry Harlow.[58] Intrigued by Harlow's now-infamous experiments on orphaned baby monkeys, Bowlby found the experimental validation for his theories that he desperately wanted. Harlow challenged the same focus on infant feeding as Bowlby when he built wire-monkey-mothers in order to demonstrate that comfort trumped food in the infant-needs hierarchy. In one particular series of experiments, he equipped bare, wire-mothers with bottles from which the infant monkeys could feed. He also placed in the cage a wire-mother covered in soft terry cloth. These soft mothers could provide comfort and warmth, but they could not provide the baby monkeys with nourishment. Through a series of tests designed to measure the monkey's behavior under different stressors, Harlow concluded that the baby monkeys sought out the soft wire-mothers for comfort and safety, but formed no such bond with the food-bearing wire mothers.[59] For Harlow, and for Bowlby, this work offered compelling evidence that feeding was superseded by other, more important instinctual needs in the infant. Harlow used his initial findings to suggest a new model of attachment, one that placed love and affection at the center of social development. "Certainly, man cannot live by milk alone," he compellingly wrote. "Love is an emotion that does not need to be bottle- or spoon-fed, and we may be sure that there is nothing to be gained by giving lip service to love."[60] Having worked with monkeys for some time on learning processes, Harlow later reflected that even before he had conducted the wire-mother study, he had become "suspicious of so-called drive reduction, hunger theory." "If you work with monkeys," he added, "you automatically became suspicious because it [feeding method] didn't make any difference when you test them."[61] Harlow's experimental findings dovetailed nicely with Bowlby's theories, a fact demonstrated in the correspondence the two maintained. A year prior to Harlow's first formal presentation of the monkey-mother results at the American Psychological Association's Presidential Address in August 1958, he wrote to Bowlby to thank him for sending him a draft of his article "The Nature of the Child's Tie to His Mother," which Harlow found to be of "fundamental importance" to his work with rhesus monkeys.[62] Together, the two helped unite the methods, theories, and evidence from the worlds of human and animal behavioral studies, providing a compelling scientific framework for understanding human mothers and infants in terms of the animal kingdom.

While Harlow and Bowlby ultimately argued that infant feeding was actually far less important than earlier theorists had assumed, the model of instinctual responses they helped create had the unintended effect of strengthening the ideology of natural motherhood, and ultimately, the case for the superiority of breastfeeding. Both psy-entists contributed to a body of research and theory that emphasized the irreplaceable psy-

Harry F. Harlow with baby monkey and cloth mother. (University of Wisconsin–Madison Archives, # S01464.)

chological and biological role of the mother in the life of the infant. In light of their work, the physiological process of breastfeeding acquired a new and far more important layer of psychological meaning for both the mother and the child. Bowlby and Harlow contributed to the creation of a scientific model of mother-infant interactions that emphasized instinctual feedback, which linked biology with emotion. An infant's crying, clinging, and rooting reflexes, in other words, were a series of species-specific instinctual behaviors honed by evolutionary development to stimulate an equally instinctual set of psychological, physiological, and emotional responses from the mother. The resulting instinct-driven feedback loop allowed the infant to thrive, but also allowed the mother's mind and body to complete its own instinctual inertia. This midcentury ethological model of motherhood added a sense of evolutionary urgency to the subject of breastfeeding. While infants had been shown to thrive on bottle or breast, what of the mother herself? Did the experience of breast- or bottle-feeding change her?

Natural Motherhood Expertise and Practice in the Lives of Margaret Mead and Niles R. Newton

The scientific validity that ethologists and attachment theorists lent to the ideas of maternal and infantile reflexes and responses informed the scientific foundation for an ideology of natural motherhood. Natural motherhood supported a model of maternal behavior based on instinct and scientifically validated parallels to the natural world. It stressed the belief that maternal-infant relations fell under the control of a complex system of physiological and psychological feedback loops. The construction of this alternative maternal ideology, however, would not be complete until science popularizers helped transmit it to the public. By the mid-1950s, some psy-entists and well-educated mothers were embedding the ideology of natural motherhood into the practice of breastfeeding. Anthropologist Margaret Mead and psychologist Niles R. Newton played central roles in this process when they united the theories and evidence of men like Bowlby and Harlow with their own perspectives on motherhood, informed by cross-cultural study and their own maternal experiences. In doing so, they would help unite biological and ethological theories of maternal-infant attachment with a perspective on maternal authority and purpose reminiscent of the Victorian era.[63] Uniquely informed by modern scientific theory, natural motherhood popularizers, epitomized in Mead and Newton, saw in breastfeeding an answer to the midcentury's anxieties about the maternal role, the connection between the mind and the body, and ultimately, the declining moral authority of the modern "Mom."

Anthropology was one of the first psy-ences to begin probing questions of gender, family, motherhood, relationships, and the debate of nature versus nurture. These early strands of thought are nowhere more visible than in the work of renowned anthropologist Margaret Mead. Certainly it had been anthropology's business for some time before Mead came along to investigate the question of "nature versus nurture," but Mead helped bring the investigation of the domestic sphere into the forefront of anthropological investigation and captured the imagination and interest of the American public.[64] From her earliest work in Samoa, Mead expanded the scope of anthropological investigation beyond questions of sexual divisions of labor and began to ask about the less obvious cultural constructs of sexual and gender systems. In her subsequent work in New Guinea and Bali, Mead's analyses offered insight into the relationships between child-rearing practices and social structures. Her work highlighted the importance of the mother-child relationship in the development of personality, the structure of the family, and the nature of a society.

The overarching questions that drove Mead's research, from her beginnings in American Samoa through her forays into popular writing in the magazine *Redbook*, revolved around cultural and biological differences in human gender and sexuality.[65] Mead's own life experiences and self-awareness were intimately tied up with her interest in these questions. Lois Banner's analysis of the relationship between Mead and her friend and mentor, Ruth Benedict, suggests that Mead sought ways to reconcile many of the tensions she experienced as an American woman, and a scientist, with her work in the field.[66] Mead questioned rigid definitions of sexuality and women's roles throughout her life. She cultivated romantic and sexual relationships with both men and women, subscribed to a philosophy of free love, and challenged the norms of compulsory heterosexuality and monogamy. Simultaneously, Mead sought throughout her life to maintain a distinctly feminine persona that often complicated her presence in a professional realm traditionally dominated by men. In both her life and her work, Mead helped reconcile older ideals of female maternal authority and respect with modern and more fluid scientific models of human culture and behavior.

After giving birth to her daughter in 1939, Mead experienced her femaleness in a way that she believed transcended cultural borders and the limiting models of female sexuality that dominated American life.[67] Through her fieldwork and her own maternal experiences, Mead developed a vision of womanhood that revolved around the biological processes of pregnancy, birth, and child rearing, with heterosexual relationships playing only one part in a much more expansive female experience. Through her work and in her own life, Mead contributed to a model of universal femininity that identified the importance of biological motherhood and the mother-child relationship as constants across cultures.[68] At a time when the strength of American families rested upon the sexual containment and obligations of women in marriage, Mead made an argument for a maternal-centric worldview that supplanted the importance of heterosexual relationships with those of mother-child bonds.[69]

Mead demonstrated her scientific and personal philosophies in both her life and her work. In a series of films she coproduced with Gregory Bateson of the Balinese (Indonesia) and the Iatmul (New Guinea), Mead explored the significance and meaning of basic differences in childbirth and infant-care practices between the two groups. Observations of the mothers' attitudes toward their infants during the mundane tasks of feeding, bathing, and comforting provided Mead with evidence that mothering, even of very young infants, initiated the newborn into a way of being that was uniquely suited to his or her place in society. Among the Balinese, for example, she observed that mothers often teased and incited jealous fits from their toddlers by fawning over the infants of their friends and family. Such a practice, she believed, was part of a functional system balanced out by her observations of the highly performative and dramatic nature of Balinese society. On the other end of the spectrum, Mead portrayed the infant-rearing techniques of the Iatmul as nonantagonistic, confident, and highly social. In her films, Mead shows the Iatmul mothers as they calmly nurse each other's babies. One mother is depicted comforting her own child even as she simultaneously breastfeeds the newborn infant of a neighbor. For comparison, Mead also depicted the Balinese mothers, who, she observes, appear to use breastfeeding as a means to trigger possessiveness and anger responses in their children. As one mother in the film is shown nursing the infant of a friend, her own toddler angrily and repeatedly pulls her breast out of the infant's mouth.[70] Mead, who was well versed in psychoanalytic theory, saw early mother-child relations as indicative of larger social phenomena. While nursing was just one way a mother interacted with her infant, it seemed to consume a significant portion of Mead's studies of family life. The fact that breastfeeding took place over a period of months, and even years, made it in many ways an ideal subject of study for anyone interested in cultural reproduction.

By 1950 Mead decided to turn her gaze homeward and she published a book on "the sexes" and modern American society. The work was critical of what she saw as America's valuation of "mechanical perfection" over the "peculiar rhythms" of nature. Having lived and experienced the "rhythms" of life in nonindustrialized cultures, Mead criticized American practices surrounding childbirth and infant care. She wrote:

> Breast feeding is frequently abandoned altogether, and by the time the child goes home, the mother, if not the baby, has learned that contacts between mother and child have certain form. . . . For the primary learning experience that is the physical prototype of the sex relationship—a complementary rela-

> tion between the body of the mother and the body of the child—is substituted a relationship between the child and an object [a bottle of formula], an object that imitates the breast.[71]

Given Mead's orientation toward motherhood, a view that highlighted a biological universalism to the female experience while also recognizing the cultural variation in specific mothering practices, it is not surprising that she held a critical attitude toward modern American practices surrounding childbirth and infant feeding. As a popular scientist and a respected cultural icon, Mead actively translated her ideals of an embodied and natural motherhood to American mothers across the country. In her popular writings and published interviews, she consistently emphasized her belief in a significant degree of universalism in the maternal experience. She was resolute in her perspective that women were evolutionarily better suited for child rearing than were men, despite the wide variation in social norms that she had witnessed in her own fieldwork. A *McCall's* interviewer described her in a 1970 article as "an idealist, she believes better forms of society could be invented, so that neither sex would be tied down to the home." Yet, in the same article, Mead claimed that she was "disturbed about fathers taking care of very young infants," a circumstance, she observed, that she had not seen in any "previous civilization."[72] Despite her doubts regarding the role of fathers, Mead generally championed a more fluid society when it came to norms of sexual behavior and family structure, beliefs she acted out in her own life with multiple marriages and female partners. In 1978, for example, she was quoted as saying, "The present nuclear family is nonviable, I believe, because it is deathly dull for a woman never to do anything creative or productive."[73] Thus Mead often challenged simplistic models of a gendered division of labor and life in both her own family and in her work.

Mead herself gave birth to her daughter in the 1930s without anesthesia and chose to breastfeed on demand, nearly a decade before such practices would become even peripherally known to the most informed of American mothers. She opposed hospital policies regarding childbirth, having witnessed many births in her fieldwork in which mothers delivered their infants while standing up and were engaged in taking care of and feeding their infants within a matter of minutes following birth. In her work and in her life, Mead helped articulate the ideology of natural motherhood as one based on a universalist and biological idea of motherhood, one which transcended cultures, time, and even species.

Mead's work articulated a woman-centric, pronatalist gender model: women could be scientists, doctors, and anything else they wanted to be, but biology was ultimately the thing that defined femaleness. When it came to love, Mead was equally pragmatic: "The infant whose world is warm, giving, and reliable responds with love that echoes the love he has received. But the infant who is continually hungry, cold and neglected will come to hate those who hurt him and do not attend to his needs. In a sense, both love and hate are learned."[74]

In Mead's world, as in her work, women were the bearers and primary caregivers of children and in that way they drove the inner workings of a society and of the family. Immersed as she was in the less technologically modern worlds of her subjects, Mead feared that Americans might lose touch with human nature itself. Her work as an anthropologist, informed by her identity as a mother, reflected a deep understanding and appreciation for the maternal experience, an experience that her own life and her time in the field had taught her to see as natural. Exposed to the processes of birth and breastfeeding in cultures across the globe, Mead concluded that there were biological continuities in the realm of motherhood despite the variations in how different cultures dealt with it. It was this articulation of an embodied maternal nature and universal identity that spoke to and inspired a young psychology student at Columbia University in the 1940s named Niles Rumely Newton.

Niles Rumely Newton's work played a pivotal role in marrying the ideology of natural motherhood with the practice of breastfeeding. It was upon her work that America's earliest back-to-the-breast movement arguably took root and flourished.[75] The narrative and analysis of her life and work reveals that her own experiences as a mother informed her study of the mind-body connection in maternal processes and drove her to ask questions few others were considering in the years surrounding World War II.

Born Niles Polk Rumely, in 1923, the youngest of four children, she began her life in the city of New York, where her father ran a newspaper. Her East Coast beginnings belie her family's roots in La Porte, Indiana, where her "great, great grandfather Polk" had helped to build the first road through Indiana from the Ohio River to Lake Michigan. Her father, Edward, was a passionate and busy man; the grandson of Meinrad Rumely, a German immigrant and the founder of a successful Indiana-based tractor company. In the course of Edward's life, he ran the family business in La Porte; established the Interlaken School for boys in

Rolling Prairie, Indiana; published and edited the *New York Evening Mail*; helped introduce dietary supplements to the American market in the form of "Vegex"; and founded the Committee for the Nation for Rebuilding Purchasing Power and Prices. He eventually returned to his roots, spending his final years in La Porte battling cancer and advocating for the promotion of cancer screening and the banning of cigarette smoking. Newton's mother, Fanny Scott, who could trace her lineage back to the *Mayflower*, was a native of La Porte and a Smith College graduate. She returned to the area after college to work at Edward's school.[76] Snippets of Fanny Scott's personality revealed in family correspondences suggest that she took great pride in being first a mother, and later a grandmother, to her tall, healthy, and intelligent brood. Her husband Edward mused in family letters that Fanny was pleased that her children had always been the tallest and youngest in their kindergarten classes, an achievement he proudly attributed to Fanny's devotion to breastfeeding.[77]

Niles attended Bryn Mawr College and, while there, met and married Michael Newton, a British student studying obstetrics and gynecology at the nearby University of Pennsylvania. The two would go on to share a life and a career together as they often collaborated on projects and publications. As a PhD student at Columbia University in the late 1940s, Newton seized every opportunity to take classes with distinguished faculty in psychology and anthropology, where her burgeoning interest in the "psychology of lactation" began to evolve into her own unique research program.[78] It was in an anthropology graduate seminar that she first encountered Margaret Mead, whom Niles would make her lifelong mentor and colleague through her persistent and dedicated efforts. Over the course of her education and career, Newton focused on the psychological and physiological connections in maternal processes. Her own experiences as a child from an avidly pro-breastfeeding family and as a breastfeeding mother herself fueled her determination throughout her career. She recalled in a 1950 interview with *Time* magazine that despite her own resolve to breastfeed, her doctors had insisted that she would not be able to nurse her firstborn. "It made me so mad," she remembered, "that I insisted on trying it. And I succeeded."[79] The experience galvanized her and set her on a path that would clarify and reorient decades' worth of disparate scientific knowledge about motherhood and breastfeeding and bring together work from the fields of ethology, psychology, anthropology, and medicine.

Niles and Michael Newton. (The University of Pennsylvania Archives, Alumni Records Collection, Box 1917.)

In 1947 and 1948, while still completing her master's degree at Columbia, Niles and Michael performed a series of experiments that would help lay the foundation for her work on lactation, motherhood, and female sexuality when they attempted to locate the existence of a milk "let down" reflex in human females. Veterinary research funded by the dairy industry had led scientists to discover that milk-ejection reflexes stimulated by sex hormones were a key part of the lactation process in animals. Furthermore, animal experimentation had revealed that hormones associated with a variety of physiological sexual responses were involved in milk ejection. The Newtons sought to understand the workings of this mysterious process in humans, or more specifically, in Niles herself. While breastfeeding her second daughter, Niles exposed herself to unpleasant interruptions to demonstrate the impact of stress on the ability of the breasts to release milk. The interruptions Niles underwent ranged in severity from the extreme (receiving a small electric shock or plunging her foot into a bucket of ice water) to the mundane (solving math problems). Even minor interferences produced similar results, slowing or even stopping her milk. To demonstrate that the sex hormone oxytocin was triggering the let-down process, Niles received an injection of this substance and was again exposed to interruptions of varying kinds and degrees. With the hormone in her system, her milk flowed regardless of the distraction. The Newtons published their findings in a 1948 issue of *Pediatrics*, the official publication of the American Academy of Pediatrics.[80] In later years, she went on to author similar work demonstrating the links between emotions, physiology, and motherhood in high-profile journals such as *Pediatrics*, the *Journal of Pediatrics*, the *American Journal of Medical Science*, and *Marriage and Family Living*—sometimes with her husband as coauthor, sometimes on her own.[81]

In her PhD work and first book, Newton expanded upon her early interest in the connections between mind, body, and culture in maternal processes.[82] Through multiple interviews with 246 mothers of newborn babies, all of whom had participated in what she characterized as "natural childbirth" in Philadelphia between 1950 and 1951, Newton argued that motherhood was a complicated identity composed of biological, psychological, and cultural components. Motherhood, she argued, was just one part of a larger female identity that encompassed sexuality and other physiological processes such as menarche and menopause, all of which were equally steeped in cultural meaning. Newton fervently argued that biological experiences were intimately connected to a woman's psychological and emotional state (and vice versa) and that both the biological and psychological experiences of motherhood were heavily influenced by the cultural expectations into which a woman was born and lived.

Beyond this web of biology, culture, and psychology, however, Newton maintained a belief that there existed a superior "nature," an innate knowledge embedded within the physical body itself, that if read properly, could indicate the path to "better" and "more natural" living. Furthermore, Newton's work reveals an underlying belief that certain biological processes both necessitated and recalled a connection to this universal human nature, particularly those pertaining to motherhood. In a paper she coauthored with Margaret Mead in the early 1960s, the two argued, "There is no such thing as natural childbirth if we mean by that term physiological childbirth without any [cultural] overlays." They went on, however, to nod at the existence of some forms of instinctual, or innate, bodily knowledge that were more *natural* than not, writing: "The only thing that can be done," to improve the birth and postpartum processes for women, "is to foster selectively those [cultural] patterns that seem more in tune with physiological cues and play down those patterns that are likely to interfere with smooth physiological functioning."[83] Newton sought to identify the fundamental biological processes of motherhood because she believed that some cultural practices were more supportive of the maternal body's innate knowledge than others. These purified maternal processes in her work were not meant to serve as blueprints, but instead Newton hoped the pursuit of identifying the truth of this "natural motherhood" would help better align the culturally mediated experiences of motherhood with the maternal body's instinctual needs. Identifying the existence of a natural, innate, and universal motherhood captivated Newton throughout much of her life and her work.

Between 1961 and 1962, Newton continued to work in partnership with Mead from her home in the Department of Obstetrics at the University of Mississippi Medical Center in Jackson. With Mead based in New York, Newton seems to have borne the brunt of the responsibility for maintaining their relationship long-distance. Despite the logistical difficulty, the two managed to produce a group of papers and documentary films to share at the First International Congress of Psychosomatic Medicine and Childbirth to be held in Paris in July 1962. The Paris trip put Newton in touch with an international network of over four hundred scholars, nurses, midwives, and physicians who shared an interest in how the biological processes of motherhood were shaped, experienced, and given meaning by cultural and psychological influences. For Newton the conference was stimulating and provocative, particularly the informal conversations that unfolded over dinner in the evenings. She recalled one such evening when the discussion turned to "the psychoanalytic hypothesis . . . that closeness between mother and newborn infant is important. Dr. René Spitz, famous for his work on mother-infant attachment, proceeded to try to prove it by citing the Harlow work with monkeys raised on wire and cloth mothers."[84]

In her expanding postdoctoral research, Newton suggested that there would be both immediate and long-term implications for individuals, families, and the culture in general if and when the natural maternal process of breastfeeding was completely abandoned. In 1967 in a lengthy discussion of what she referred to as the "transition period," the period after birth but prior to infant autonomy, across over two hundred cultures, Newton argued that in the United States the transition period was "patterned so that it virtually ceases to exist."[85] She wrote, "Since modern technology has not yet invented a feasible substitute for gestation, pregnancy and childbirth cannot easily be muted, but substitutes for lactation have been invented. . . . The transition period of physiological dependence can be, and often is, eliminated."[86] She pointed to Harlow's work, among others, to suggest that maternal behavior during "the transition period may crucially influence subsequent behavior" of the mother. Newton went on to make the case, again, that despite the cultural differences seen from place to place and people to people, there was enough evidence to suggest that "close mother-baby contact, sensitivity to crying, child spacing, and prolonged breast feeding occur together again and again" because "there may be strong mechanisms involved in this interrelation of patterning."[87] In other words, the renowned Dr. Benjamin Spock's reigning advice of the day, to let a baby "fuss or cry for fifteen or twenty minutes or more" before feeding him, was part of a larger system in which "compulsory separation of mother and infant" immediately after birth in most hospitals fed easily into the practice of formula feeding rather than breastfeeding.[88] Western cultural patterns divorced women from their instinctual maternal processes, argued Newton, and this was an unfortunate thing, most certainly for the child, but even more importantly for the mother herself. It was mothers whom Newton depicted time and time again as victims of a modern American culture that devalued women on dual fronts, both in the workplace and at home. Newton's scientific and popular work emphasized that women could learn to listen to and support their maternal instincts. This was one way, she argued, that women might reclaim a sense of self-worth and purpose, things that she believed women were generally lacking in midcentury America.

From early in her career, reaching America's undervalued mothers was a primary professional goal for Newton and as a result she began cultivating a popular scientific presence for herself early on. In 1957 she published a parenting advice manual, which went through twelve printings, the last of which was issued in 1981. Newton's advice book graced the "suggested readings" list in La Leche League's hugely influential book *The Womanly Art of Breastfeeding*, first published in loose-leaf form in 1958. They recommended Newton above all other advice books of the day, quite a high appraisal given the fame of other child care experts of this era.[89] Newton's work helped add not only knowledge but also legitimacy to the League's efforts in challenging the hegemony of a bottle-feeding culture. In return, Newton and her family remained deeply connected to the organization for decades. In 1962, Newton wrote to Mead about her personal involvement with the founding of a League chapter in Jackson, Mississippi, where the Newtons spent a decade of their lives. Niles was profoundly excited about the camaraderie and support the League provided, a personal experience she later recalled in a speech she gave at the 1960 National Convention for Childbirth Education in Milwaukee, Wisconsin. "There were nine breast fed babies at that meeting," she remembered, "and many of them got discretely [*sic*] nursed." She added, "For the first time in my many years of breast feeding four children I didn't feel odd—I felt I was in a group where breast feeding was natural, the customary thing to do."[90]

Many mothers recognized Newton's contributions to their own evolving sense of self and they wrote to express their gratitude. In a 1967 letter to La Leche League one mother gushed: "Just had to write to tell you how much I'm enjoying Dr. Newton's book. I've already read the first 16 chapters and plan on starting all over again when I've finished. She has such a wonderful approach to every aspect of child care. I've thoroughly enjoyed every page."[91] Newton actively cultivated her public persona, becoming a contributing editor for *Baby Talk* magazine and later writing a weekly column called "Your Family, Your Child," for the *Chicago Tribune*. Newton's popular works mirrored the messages of her academic publications. Her writings stressed both the physical and emotional importance of the mother-baby relationship but she also increasingly advocated for greater partnership between mothers and fathers in the home and in the care of older children so that women could pursue interests beyond child rearing once their infants were grown.

Psy-entific Perspectives on Natural Motherhood

At the end of his 1958 address to the American Psychiatric Association, Harry Harlow famously turned his attention to the exchangeability of mothers and fathers. If infant development and attachment were uncoupled from feeding, he argued, then the logical conclusion was that "the American man" could be just as capable at raising a child as a woman. He said, only somewhat tongue in cheek: "It is cheering . . . to realize that the American male is physically endowed with all the really essential equipment to compete with the American female on equal terms in one essential activity: the rearing of infants."[92] Newton, with the perspective of a mother and a scientist, saw things differently. She wrote: "The breastfeeding mother has a real physical need for the baby just as the baby has need of the mother."[93] In the end, Newton suggested that breastfeeding may or may not do amazing things for infants, but its greatest benefit, its importance to the breastfeeding mother and her family, was something to which few mainstream experts in her time were then giving much thought.

The science of motherhood in the mid-twentieth century was engaged in a process that Rebecca Plant has described as the development of a "psychological and biological conception of the maternal role," a role that she suggests was a demotion from the "lofty spiritual connotations" of the Victorian ideology of moral motherhood.[94] What the elucidation of this ideology of natural motherhood shows, however, is that the divide between modernists and moralists was perhaps not so stark. Rather than being incommensurable, biological and moral motherhood could be, and in fact were, interwoven. This was nowhere more evident than in the life and work of Niles Rumely Newton, who helped create and popularize an ideology of natural motherhood and breastfeeding in the mid-twentieth century. Despite the attempts of her contemporaries, like Harry Harlow, who sought to uncouple the act of infant feeding from maternal-infant bonding, maternal scientists like Newton demonstrated that data in the human and animal sciences could also be wielded in support of the importance and centrality of motherhood and breastfeeding. While Harlow chose to interpret his own data on infant development as *dismissive* of the importance of a mother, by the 1950s there was a growing peripheral network of scientists, mothers, doctors, and nurses who saw breastfeeding as a unique and positive process with real psychological and physiological benefits for the mother and the child.

This stance toward infant feeding and mothering not only assumed the importance of the mother-infant relationship for psychological fulfillment in both the infant *and* the mother, but it also scrutinized every aspect of the mother's behavior for signs of maternal fitness. In this era, the smallest gestures came to have meaning for scientists and analysts. Robert R. Sears, the popular child-development specialist who oversaw the Yale Child Guidance Clinic, contributed his thoughts on the issue of infant feeding at a meeting of the American Philosophical Society in 1948. He explained that infant feeding mattered because

> mothers do not merely feed their children; they talk to them and snuggle them, they pick them up frequently or rarely, they are tense or relaxed. . . . Each mannerism, each approach or withdrawal, each facial expression, laughter and frowning, each command or plea, is a quality of the social person, beta, in the interaction sequence. These qualities, to the extent that they occur with reasonable consistency in the parental interaction with the child, become essential properties of the goal object or social person around whose behavior the child's secondary motives are developed.[95]

The significance of this perspective toward motherhood at midcentury is only fully realized when examined within the context of broader shifts in scientific and cultural ideas about motherhood, maternal instinct, and infant development, as this chapter has highlighted. While Freudians had long been concerned with feeding and the process and timing of weaning, newer ways of thinking about motherhood and infant feeding emerged across disciplines, with psy-entists like Bowlby, Harlow, and Ribble leading the way. Their work, combined with maternal psy-entists and popularizers like Mead and Newton, helped to construct an ideology of natural motherhood that was deeply connected to breastfeeding and responsive to popular anxieties regarding gender, sex, and family. While it was true that the work of researchers like Bowlby and Harlow allowed for more leeway when it came to the breast versus bottle decision, by emphasizing instinct, "the natural," and an embodied identity of motherhood, these psy-entists succeeded primarily in changing the terms of the debate. Despite the cultural resonance of the ideology of natural motherhood in the postwar years, mothering "naturally" did not come easily. Not only did mothers lack access to adequate information and support when it came to breastfeeding; they encountered a host of obstacles from the medical system. In this period the medicalization of childbirth and childhood, more generally, brought more and more mothers and their infants under the authority and advice of physicians and a growing medical system.

CHAPTER TWO

Frustration and Failure

The Scientific Management of Breastfeeding

In 1942 in a Philadelphia row house, a young Catholic mother settled back home after the birth of her second child, whom she intended to breastfeed, just as she had with her firstborn in 1937. Already an experienced mother, she went about her daily tasks believing she was doing everything right for the first few weeks. Her elder daughter, Barbara, recalled what happened next: "We were poor and went to a clinic and the nurse would show up and weigh the baby once a month. And when she came my sister had lost a pound of birth weight."[1] Upon learning she did not "have enough milk," Barbara's mother panicked and with the encouragement of the nurse switched her baby to formula and solid foods. This scenario repeated itself with a few additional twists less than two decades later when Barbara set out to breastfeed her first baby in 1960. After giving birth in an "inner-city hospital" in Philadelphia, Barbara recalled that she "told them that I wanted to breastfeed . . . and they said fine, fine." Despite her wishes, she later found out that the nurses had been giving her "dry-up pills" without her knowledge. By the time Barbara left the hospital five days later, she had been unknowingly taking lactation suppressants for almost a week. Despite this hiccup, she continued to try to breastfeed once she returned home. There too, however, she encountered a lack of support, this time from her pediatrician who responded with an audible groan upon learning that his patient was set on breastfeeding. Six weeks later, after undergoing a mild outpatient medical procedure, her milk production decreased. Fearing she was losing her milk just as her mother had before her, she switched to a formula and ended her first breastfeeding experience.[2]

The stories of Barbara and her mother are not unique. Women of varying backgrounds chose breastfeeding during the 1940s, 1950s, and 1960s despite the realities of an increasingly unsupportive medical culture consisting of general practitioners, obstetricians, pediatricians, nurses, and hospitals. In doing so, they faced the institutional inertia and authority of orthodox medicine and the ideology of scientific motherhood that had emerged since the late nineteenth century. Scientific motherhood at midcentury encompassed an interventionist, medical- and technology-driven model of maternal and child care that rested upon the expertise of scientifically trained professionals as authorities and guides. As the appendages of the medical system, health care providers approached their patient interactions from a variety of vantage points, all of which seemed incommensurable with the emergence of an ideology of natural motherhood and the breastfeeding model it supported.

Scholars who have explored the history of breastfeeding's decline have argued that bottle-feeding under the guidance of a physician was widely accepted as "normal" by the 1930s and 1940s. Even though doctors throughout this period often maintained the mantra "breast is best," few believed that the breast was the normal or even most desirable option for the majority of mothers. The existing historical narrative indicates that few if any mothers questioned the effectiveness and efficiency of the rise of scientifically managed bottle-feeding throughout the postwar years.[3] In fact, as Jacqueline H. Wolf has convincingly shown, mothers' own concerns over "insufficient milk" helped secure the place of the bottle in American infant feeding by the 1910s.[4] While such accounts offer critical insight into the complex forces at play in infant feeding shifts, they have also tended to obscure the meaningful persistence by some segments of the medical and scientific community in their support of breastfeeding. In other words, the answer to the question "Was medical advice for or against breastfeeding in the postwar decades?" is not as straightforward in this period as we might believe. As historians have pointed out, doctors throughout the postwar decades did often pay lip service to the idea that "breast was best" even while they believed most women were incapable of doing it. What we have yet to better understand, however, is what those few physicians who were committed to breastfeeding were doing, why they were doing it, and what impact their efforts ultimately had (or didn't have) on the late twentieth-century resurgence in breastfeeding.[5]

An investigation into the state of scientific knowledge about breastfeeding throughout the mid-twentieth century furthers our understanding of why and how physicians managed their breastfeeding patients the way that they did. When physicians sought information about how to scientifically manage a breastfeeding mother in the postwar years, they faced a frustrating dearth of readily available information. Although published research on lactation did exist, it was scattered across a multitude of disciplines and much of it was based on animal studies rather than observations of humans. Nevertheless, doctors and researchers who remained committed to developing a more legitimate science of breastfeeding pushed forward and sought ways of utilizing existing research and observations drawn from a broad range of fields to advocate for better care of nursing mothers and babies. At the same time, the growing association between breastfeeding and the ideology of natural motherhood throughout the postwar era made physician oversight of breastfeeding less and less tenable. Doctors who continued to support breastfeeding up through the 1960s increasingly came up against the realities (whether they realized it or not) that breastfeeding demanded individualized care, a holistic view of health and the body, and a degree of flexibility that modern medicine and the ideology of scientific motherhood it supported were ill equipped to provide.

The State of Breastfeeding Science at Midcentury

Throughout much of the mid-twentieth century the medical community struggled in vain to come to a consensus on the issue of breastfeeding. It was not until the late 1970s that the American Academy of Pediatrics made any official statement explicitly in support of breastfeeding as the "only source of nutrients for the first four to six months for most infants."[6] Despite their own complicated relationships to infant feeding, throughout the 1940s, 1950s, and 1960s, certain pediatricians, obstetricians, general practitioners, and nurses observed signs of a renewed interest in breastfeeding among some of their patients, even as the rates experienced an overall decline.[7] A widely cited report published by Dr. Katherine Bain of the U.S. Children's Bureau on the diminishing incidence of breastfeeding in 1948 prompted years of discussion in medical journals and at meetings about whether breastfeeding was on its way out and what, if anything, ought to be done about it. Many physicians would come to see Bain's report as confirmation of their own day-to-day clinical experiences.[8] As Bain's numbers showed, fewer and fewer American mothers breastfed at the time of discharge from the hospital, and even fewer breastfed their infants for longer than six weeks. And yet, as early as the 1940s, there were some physicians and nurses who also began to comment on the possibility of a breastfeeding resurgence. As Dr. Claude Heaton, a New York obstetrician observed, 73 percent of his patients were breastfeeding one day postpartum, but by hospital discharge less than a week later, he saw only about 66 percent doing so. Observations like these along with Bain's study raised difficult questions: if so many mothers started out breastfeeding, why were they failing to maintain it in such large numbers and who was to blame for this trend? Clinical observations led Heaton and other like-minded physicians, to conclude that "a fairly high percentage of mothers would still like to nurse their babies."[9] Data like this prompted lively discussion on the subject throughout small pockets of the medical community, even as practices supporting scientific feeding continued to grow more deeply entrenched.

In response to the decreasing breastfeeding rate, the documented cases of breastfeeding failure, and the mothers who continued to show up who wanted to try breastfeeding, health care providers turned to the medical literature to see what was available on the subject only to find a complex and disparate knowledge landscape. Research on breastfeeding by the 1950s was characterized on the whole by small-scale animal-based studies and clinical case observations, but one also had to look to psychological investigations and research published beyond the United States in order to fully assess the state of the field, such as it was. For many physicians the lack of a coherent body of literature on breastfeeding made the studies that did exist seem circumstantial and suggestive, at best. As the Boston pediatrician Dr. Clement A. Smith put it, "conscientious physicians, anxious to give sound scientific advice, find themselves falling back among traditions and impressions. . . . Obviously we need facts."[10] Medical science had devoted its investigations of human milk to re-creating it in a lab for decades with relative success, but on the whole it had mustered little organized or sustained effort toward understanding the actual processes involved in human lactation.

Faced with a lack of contemporary work on the subject, physicians turned to research done in the early decades of the twentieth century for answers about breastfeeding. Many in this position would have come across an article by Dr. Clifford H. Grulee from Chicago's Rush Medical College. Published in 1934 in the *Journal of the American Medical Asso-*

ciation, Grulee's work was some of the only accepted scientific evidence that breastfeeding actually was better than formula. Grulee's study consisted of data collected from the records of 20,061 babies under the care of the Infant Welfare Society of Chicago between 1924 and 1929. Of this group of largely poor and immigrant babies, 48.5 percent were breastfed for nine months, 43 percent received mixed feedings of breast milk and formula, and 8.5 percent were artificially fed. The doctors used data from these groups to compare infection rates, particularly gastrointestinal and respiratory diseases. They reported, "The total morbidity of the breast fed group was 37.4%, of the partially breast fed group 53.8%, and of the artificially fed group 63.6%." The mortality figures were low overall (1.5 percent), but even more dramatic differences appeared in correlation with feeding method. Of the 218 deaths that occurred during the study period, 6.7 percent were in the breastfed group, 27.2 percent were in the partially breastfed group, while 66.1 percent of the dead had been completely artificially fed.[11] The same team published a similar report in a 1936 issue of the *Journal of Pediatrics*, which also surfaced in printed breastfeeding discussions in the early 1950s.[12] Another study published in 1936 by Grulee and his colleague Heyworth N. Sanford, medical director of the Infant Welfare Society, argued that infants fed only breast milk also had lower incidences of eczema.[13]

While breastfeeding supporters continued to cite this data as evidence of the natural superiority of human milk, by the end of the 1940s health care practitioners overall were so confident in the advances in sterile formula feeding that most looked to data like Grulee's as a thing of the past. As one article put it, "A few years ago the incidence of mortality among artificially fed infants from bacterial contamination was much higher than that among breast fed babies, but at the present time there is no distinct difference."[14] In fact, statements like the former, although common and uncontroversial at the time, were scientifically unfounded. The reality was that doctors who looked for hard data could find little distinction between artificial feeding and breastfeeding because researchers had failed to study breastfeeding in any significant way beyond merely cataloguing its chemical content. Research comparing short-term health outcomes in infants such as Grulee's looked increasingly outdated when viewed through the lens of modern postwar medicine. Antibiotics could treat infections, intravenous fluids could save infants from dehydration, and infants could appear to thrive on artificial formulas as their weight gains could often match or outpace those in breastfed babies.[15] Because of these advances in the safety of artificial feeding, doctors who saw babies who did not appear to be nursing successfully, whose mothers encountered challenges along the way (such as sore nipples, engorgement, and abscess), increasingly saw formula as the obvious and easy answer. It was for all of these reasons that British pediatrician Dr. Harold Waller could state in 1946 with conviction that "in the last fifty years, which have seen so much achieved in the protection of infant life, little has been added to our understanding of the unreliability of lactation."[16] By the early 1950s, however, there were those in the health care professions who began to question this dearth in understanding.

Some of these internal critiques came from physicians who had long advocated breastfeeding. These were doctors who had worked through the days "when dirty milk was the rule rather than the exception and when breast-fed babies in New York tenements had better chances for life and health than their bottle-fed contemporaries in Fifth Avenue."[17] Their experiences made them wary of the trend toward artificial feeding as the rule. Such was the case with Dr. Frank Richardson, from Asheville, North Carolina, who recalled, "Until 1919 breast feeding was looked on by the average physician as a providential occurrence." He wryly noted that "everybody talked about breast feeding but no one did anything about it." In 1919, he remembered, a physician named J. P. Sedgwick in Minneapolis began an intensive breastfeeding program that taught mothers how to manually express breast milk and impressed upon his medical students, physicians, mothers, and nurses alike that "almost every mother can nurse her baby."[18] Sedgwick reported overwhelming success in his program, claiming that he had 96 percent of his patients breastfeeding successfully two months after delivery, and 72 percent after nine months.[19] Despite Sedgwick's success with his Minnesota clientele, many physicians balked at instituting similar programs, claiming that his results had been due to the fact that he worked "with sturdy women not far removed from the farm and that sophisticated Eastern city women would never be able to give similar high percentages of breast feeding."[20] Richardson himself, however, begged to differ, claiming that "a similar demonstration was put on in Nassau County, a suburban section just outside the city of New York," and that they were able to achieve results "that were almost identical."[21] Richardson also pointed to the 1923 work of "a pediatrician who had been a farm boy," M. L. Turner, who compared his experiences and knowledge of cow milking and management to that of lactating mothers, suggesting that physicians who had become too far removed from the nature of the pastoral countryside were working at a disadvantage when it came to overseeing breastfeeding mothers.[22] Conversely, when it came to a scientific understanding of the processes involved in human breastfeeding, Richardson compared the lack of a legitimate body of literature on the subject to the early days of the twentieth century when physicians could not agree (for lack of knowledge) on the best way to feed an infant artificially. Ironically, Richardson pointed out that physicians could then, in 1950, be said to be wallowing in the same mire of ego and hearsay as their forebears half a century before them.

Richardson's scathing assessment of the field fell largely on deaf ears, as the attitude among physicians became increasingly one of disinterest in the process of breastfeeding despite the avid discussions carried on at meetings and in journals. Few physicians stepped up to fill the shoes of the aging generation of breastfeeding evangelists. Despite their scarcity, there were a few champions of breastfeeding who continued to push the study and the importance of lactation. Among these was Edith B. Jackson, head of Yale's Grace–New Haven Community Hospital's rooming-in program. American breastfeeding science also owed much to the interdisciplinary duo of Michael and Niles Newton, who published on the subject throughout the decade. In 1951 Niles and Michael Newton published a literature review on breastfeeding science, drawing information from experimental physiology, veterinary science, and dairy science as well as psychology and medicine. Even then, the Newtons were able to find strong evidence for breastfeeding's worth and, most importantly, information on how it worked. Although reviews may not have been breaking new scientific ground, the Newtons' publications responded to attitudes like that of Clement Smith, who argued that physicians were limited because there simply was no available information on breastfeeding. The Newtons surveyed a wide swath of literature from a variety of disciplines. They gleaned much of their data from such notable periodicals as *Pediatrics* and the *Journal of Pediatrics*, the *Journal of the American Medical Association*, and numerous other publications, including the *Lancet* and the *British Medical Bulletin*, in addition to lesser-known articles from publications across Europe. In an article arguing for "The Adequacy of Artificial Feeding in Infancy," for example, Harvard-based pediatrician Stuart Stevenson showed that all other things being equal, breastfed infants continued to show greater resistance to respiratory infections in the second six months of life (despite weaning at between three and four months of age) than did their formula-fed peers.[23] While Stevenson himself utilized his data to suggest vitamin inadequacies in formula, the Newtons looked instead to the evidence for the breastfed infant's superior immunity.[24] Data from a group of Italian researchers based in Rome's Istituto Superiore di Sasnità provided early histological evidence in 1948 that breastfeeding conveyed immunological benefits when they reported that a "lysozyme is present in human milk and colostrum but not in cow's or sheep's milk." It was a finding both the authors and the Newtons found "to be of great importance for the development of proper flora in the infant's gastrointestinal tract."[25] By the mid-1950s, researchers had found that certain gastrointestinal flora flourished only in the digestive systems of breastfed infants, lending further legitimacy to claims that breastfeeding delivered immunological and digestive benefits beyond basic nutrition.[26]

For details on the actual physiology of lactation, however, the Newtons and other breastfeeding supporters had to look beyond the confines of pediatric medicine. As early as the 1910s, physiologists and dairy scientists were producing data on the mechanisms involved in lactation in animals. Most often these investigations centered on dairy cows,[27] but studies of goats,[28] dogs,[29] cats,[30] rats,[31] and even the occasional human filled out the literature on lactation in animals.[32] One of the studies cited most frequently in the medical literature on breastfeeding in the late 1940s and 1950s came out of a research team at the Kentucky and Minnesota Agricultural Experiment Stations in 1941. Fordyce Ely and W. E. Petersen were among the first to firmly distinguish two physiological processes involved in lactation—"the synthesis or secretion of milk within the gland, and the act of ejecting the milk from the alveoli and the small ductules," a process they dubbed the "let down."[33] The dual nature of the lactation process is today understood to be a key component of breastfeeding success, because the *flow* of the milk out of the breast is physiologically distinct from the body's *production* of milk by glands in the breast. The majority of women produce milk within a few days after the delivery of a baby, but achieving the actual flow of milk out of the breast is where technique and expertise is most needed.[34] This early lactation research grew out of studies of mind-body connections, particularly the action of hormones released by the pituitary and adrenal glands.[35] Numerous investigators experimented in the early decades of the twentieth century with the impact of pituitary gland extract, or "pituitrin," injections on the lactation process in a laboratory setting. Physiologists learned early on from these studies that the extract seemed to have both a stimulating effect on smooth muscle and demonstrated "pressor activity" or an "ability to raise the blood pressure." It was not until 1927, however, that a team of researchers from the laboratories of Parke, Davis and Co. successfully separated the substances from the pituitary gland into two distinct substances, oxytocin (which was found to stimulate uterine contractions and other smooth muscle motion) and what lead author Oliver Kamm termed "pressor" (now more commonly referred to as vasopressin).[36]

By the 1930s and 1940s, Ely and Petersen's research could also build on anatomical knowledge that the smooth musculature found around "the small ducts and alveoli, and the act of milk ejection consists in the contraction of these muscles."[37] With injections of "pitocin" (oxytocin), therefore, the researchers hypothesized that the ejection mechanism or "let down" in the udder of a cow would improve, while injections of the hormone adrenaline (produced naturally in the adrenal gland), meant to mimic the exposure of a stressor or "fright," would inhibit let down. The results of their rather ingenious and elaborate investigation (involving dairy cows, cats, and exploding paper bags) led them to the conclusion that "the positive act of 'letting down' milk may best be explained as a conditioned reflex, and directly due to a high intra-glandular pressure caused by the presence of active oxytocin in the blood, which is responsible for the contraction of the alveoli and small ductule musculature."[38] "On the other hand," they noted "the failure to 'let down' milk is similarly due to the presence of adrenalin in the blood, which prevents the muscular contractions which are responsible for the high intra-glandular pressure."[39] With the publication of Ely and Petersen's work, therefore, the medical community had good reason to believe that lactation in the human mother, like the cow, depended upon a complex balance of hormones in her system at the time of breastfeeding. Stressful situations and negative emotions were likely to cause the body to release adrenaline, causing an abrupt halt to the body's let down, with the end result being a mother with breasts filled with milk but no apparent ability to feed her child. Michael and Niles Newton expanded upon the work of Ely and Petersen and confirmed that this process did work similarly in humans. Utilizing Niles herself as a research subject, they showed that her body responded to fright and stress by discontinuing the let-down reflex, and responded to injections of oxytocin by letting her milk flow freely regardless of the interruptions visited upon her.[40]

Doctors who remained committed to breastfeeding relied upon experimental studies like these through the 1950s and 1960s to better manage their patients, to argue against hospital practices and health care interventions that hindered this complex psychophysiological process, and to develop an ever greater understanding of the factors involved in a successful breastfeeding experience.[41] The overall lack of research on human mothers, however, led many in the medical community to remain unconvinced (or altogether unaware) that there were any useful parallels to be drawn between animal studies and humans. In fact, studies that showed the physiological connection between emotions and breastfeeding often seemed to bolster the belief that modern American women, too stressed by the demands of civilization, were all the more incapable of breastfeeding. Rather than stress them further, the reasoning went, "physicians should be realistic and accept the fact that the large majority of . . . infants will not be breast fed."[42]

Despite this kind of rhetoric and a handful of concerned and outspoken advocates in the medical community, women continued to be subjected to a host of stressors in the hospital, particularly during labor and delivery. Commonsense wisdom and experience, as well as science, told many conscientious physicians that a woman traumatized or subjected to external stress during childbirth faced greater difficulties with infant care in the hours, days, and even months after the birth. In a study of the effects of prenatal education and maternal care on the outcome of mother-infant welfare at the prenatal clinics of the Grace–New Haven Community Hospital, Augusta Stuart Clay reported that stress, fear, and anxiety in mothers resulted in negative outcomes during and after labor and delivery. Many of the mothers and fathers in her study reported mounting fear and anxiety leading up to the birth in the final months of pregnancy, a state compounded by their doctors' disinterest in their personal lives. In the initial hours after birth, the first feeding experience could aggravate what had already been a difficult and frightening time. "Most mothers asked the consultant [from the study] for help when breastfeeding and for specific instructions; said they felt frightened; wanted help holding the baby," Clay observed. Additionally, the study noted that mothers who had received "opiates" had babies who were "drowsy," those given cathartics had babies with "loose stools," and that when the baby cried at first feeding mothers were "quick to say her milk was inadequate" and were easily "discouraged if [the baby] did not wake or refused her breast."[43] Mothers expressed overwhelming desire to remain with their partners throughout the birth experience, and afterward. "Mothers who were breastfeeding," she reported "wanted their husbands with them at nursing time," something not normally allowed at the hospital. "Finally, in a few cases," she wrote, "when the mother was in a private room the father was permitted to scrub, put on a doctor's coat and visit. His interest and encouragement stimulated success."[44] These findings provided clinical data that again suggested just how important emotions and personal experience were in the birth and breastfeeding process for mothers and infants. Clay's study also reported that "at the close of the study, medical and psychologic examinations indicated that the breastfed babies seemed to be the healthiest and happiest," regardless of whether or not the birth had been complicated. And yet, despite provocative research like this, the majority of hospitals would maintain procedures that resisted the wishes of mothers to have a companion with them during labor and delivery, which hindered their efforts to breastfeed rather than supported them, and that kept the processes of labor, delivery, and postpartum care cold and impersonal for decades.[45] Despite observations by doctors, researchers, and mothers alike, for example, that sedation during childbirth impacted the newborn infant after birth, it would not be until the 1960s that researchers produced the laboratory results necessary to begin garnering widespread medical attention.[46]

If clinical and laboratory research by people like Waller, Richardson, the Newtons, Clay, and others explained the benefits of helping mothers breastfeed and the keys to promoting the process successfully, then why did nothing seem to change about the overall approach to maternal and infant care? These studies and discussions appeared in elite medical journals throughout the mid-twentieth century and yet throughout the 1960s rates of breastfeeding continued to fall as policies and practices remained largely unchanged. The majority of these articles, however widely circulated they may have been, received relatively little scholarly attention at the time. Newton and other breastfeeding supporters in the scientific and medical community struggled to translate the worth and value of breastfeeding to a body of experts dedicated to the tenets of scientific motherhood and a vision of the physician as an active manager of the maternal experience. Experts like Richardson, who in many ways represented an older medical mentality in which there remained a place for a sentimental "nature" in the processes of motherhood, were on the decline.[47] Experts from the psy-ences, meanwhile, began to articulate the superiority of natural processes like breastfeeding in scientific terms and yet they continued to remain on the periphery of mainstream medical practice, a minority who believed that breastfeeding should be the control, as opposed to the variable.

Placing this story into the proper context requires some understanding of the history of medical specialization, particularly in pediatrics. By the end of World War II, pediatrics as a field was becoming home to a series of subspecialties, including endocrinology, gastroenterology, oncology, and several others.[48] Simultaneously, as emblemized by the work of Dr. Benjamin Spock, a push for the pediatric generalist emerged to compete with this segmentation movement. C. Anderson Aldrich (Spock's mentor), Spock, and T. Berry Brazelton, among others, pushed to incorporate the holistic, mind-body perspective of the psy-ences into pediatrics and, in the process, adopted a perspective of patient care that resisted the reductionism of subspecialization.[49] Many in the medical community, including in pediatrics, however, dismissed the science of emotion, behavior, and the connections between mind and body as too ethereal and sentimental. Pediatric training, and medical education in general throughout this time, continued to focus on pathology, despite a growing sectional interest within pediatric (and obstetric) care toward a more psychosocial care model. This meant that while many within the pediatric community lamented the lack of general understanding of basic well-patient care practices (like breastfeeding) in the profession, the vast majority of rank and file received specialized training that stressed pathology and laboratory-based medicine. Through much of the second half of the twentieth century, in fact, it remained possible in most pediatric residency programs to graduate to practicing medicine without ever seeing a mother breastfeed an infant.[50] By the late 1940s, the observation of Hilde Bruch and Donovan McCune that "the profound importance of emotional influences for the healthy or abnormal development of children is now generally recognized" was both true and contested.[51] There remained many in the medical community who accused psy-ence-minded researchers and practitioners of overstressing the role of emotions and of being too "soft" in their approach to medicine.[52] Physicians frequently criticized the discourse surrounding breastfeeding for being "too emotional" and wondered if the fact that so many women "seemed to feel guilty and unhappy" when they did not nurse was more harmful than the benefits bestowed when they achieved breastfeeding success.[53] As historian Sydney Halpern has observed, in the 1950s "pediatrics was endeavoring to raise its standing within medical schools by promoting the scientifically grounded subspecialties."[54] Faced with pressures from within the medical profession (in 1965, pediatricians were outnumbered by general and family practitioners by more than four times), there were those in pediatrics who felt their authority threatened from within their own ranks, from those who wished to look beyond the legitimate confines of scientific medicine.[55] In fact, it seems more than noteworthy that a significant proportion, and perhaps even a majority, of the researchers who produced studies about the importance of things like prenatal education and attention to maternal needs and wants (including breastfeeding) were women. While key members of the professional elite helped "the new pediatrics" take root, the psychosocial perspective and research that supported efforts to integrate medicine with breastfeeding remained at the professional margins—associated with women, with sentiment, and outdated folk knowledge.[56]

Because the medical community at large did not entirely accept the legitimacy and authority of breastfeeding researchers, their work, while not discounted altogether, remained highly contested. Those who advocated a psychosocial approach to infant care had to be careful to walk a fine line between advocating for the importance of emotion in the care of patients and maintaining a "rational" perspective. Advocates of breastfeeding, demand feeding, and rooming-in were expected to remain objective on such matters, and to "protect parents . . . against impossible emotional demands and the psychologic double-talk that 'Heaven will fall if you do and earth comes to an end if you don't'."[57] If they came across as "too emotional," such advocates ran the risk of being discounted outright for being ideological and unscientific. In a treatise opposing the rise of "demand feeding," the system by which an infant is fed as per his or her hunger cries, a Massachusetts obstetrician lambasted a group he identified as "child psychologists" who maintained "that a new-born infant should not be nursed at regular intervals but whenever it cries; otherwise, they say, it will suffer in later life a sense of frustration."[58] "How they know this," he added pointedly, "they do not say."[59] Others looked upon the encroachment of psychology with less levity. Obstetrician Phillip Rothman, a physician from Los Angeles, California, believed the "enthusiasm for demand feeding," to be "almost incredible," and blamed it on what he believed was "an ever growing interest in psychiatry and an unfounded optimism in the belief that if one can get hold of the precipitating causes of emotional conflicts, the life of an individual can be shaped."[60] Rothman argued ultimately that the rise of demand feeding and the degree to which it relied on the *mother's* ability to recognize the infant's hunger cries had resulted in "a deterioration in the art of infant feeding" in medicine.[61] In these defensive denials, professional and personal insecurities become visible. Anxieties over the emergence of the "new pediatrics," as well as concerns that a mother might be able to exercise a kind of tacit knowledge beyond the control and authority of a doctor, are palpable in these kinds of pronouncements.

Disagreements over what (and whose) data counted and to what extent exhortations to breastfeed could be adequately and scientifically supported ultimately drove health care providers down the path of bickering over whether breast really was best, if there was enough reliable data to support such claims, and if it was worth convincing a mother to breastfeed if she did not relentlessly pursue the idea on her own. Despite the growing traction of natural motherhood ideology (evidenced by emergence of childbirth education, natural childbirth, rooming-in programs, demand feeding schedules and the pursuit of breastfeeding) and despite the efforts of staunch advocates of breastfeeding and labor and delivery reform, by the end of the 1950s even fewer mothers were breastfeeding at all and the overall tone of the medical literature had markedly shifted.

Managing Breastfeeding in the Age of the Bottle

In 1958, Herman F. Meyer from Northwestern University Medical School and Chicago's Children's Memorial Hospital published a follow-up study to the 1948 survey by Katherine Bain on breastfeeding incidence. Meyer's results of over nineteen hundred hospitals and 2.25 million infants confirmed what smaller studies had been grasping at for decades: breastfeeding was declining, and rapidly.[62] Whereas Bain's 1948 study had found that over 50 percent of babies discharged from hospitals were receiving some breast milk in 1946–47, Meyer reported that in 1956 only about one-third of all newborns discharged were receiving any breast milk, while the percentage of babies receiving only formula had nearly doubled, rising from 35 percent to 63 percent.[63] While there were those in the medical community who remained advocates of breastfeeding, statements like "[infant formulas are] an improvement on the natural product, for they contain extra iron, absent in breast milk, and vitamin supplements" became more common and acceptable.[64] By the end of the 1950s, physicians felt more secure than ever before in their ability to feed babies via scientific means and as rates of breastfeeding declined, mainstream medical interest in studying or advocating the practice continued to wane.[65]

Evaluations of feeding success focused primarily on mortality and morbidity statistics and infant growth. Improvements in formula safety and artificial feeding practice caused gastrointestinal disease rates to fall, and infants appeared to grow hale and hearty on their cow's milk formulas. Pronouncements declaring "no difference" between breast or bottle-feeding eventually started to give way to those who claimed formula was actually superior in some ways. Similarities between breastfed and bottle-fed infants in this period may have seemed even more blatant since even mothers who breastfed often took the advice of medical authorities and utilized supplemental and substitute feedings of formula.[66] Despite years of discussion and published clinical experience, many in the medical community insisted that the mantra "breast is best" was by then an outdated and unsupported claim. The "benefits" then ascribed to breastfeeding remained stable throughout these years and generally (with very little variation) went as follows: it's convenient, lower in cost, inherently safer, emotionally satisfying for mother and baby, and it's "natural."[67] By the 1960s, however, more and more physicians argued that the majority of these touted benefits seemed to have been matched by advances in formula preparation and eroded by clinical experience.

Boosters of modern medicine and scientific motherhood proclaimed the safety of formula, declaring it an equal to breast milk even while contemporary studies suggested this was not the case. In a study comparing two formula delivery systems in a hospital setting, the authors reported that only 92 percent of the in-house formula samples they collected were considered "satisfactory in bacterial count." Compared to the bacteriological safety of breast milk, the acceptance of an 8 percent contamination rate (not including the samples in which bacteria levels were found to be "satisfactory") seems ludicrous in retrospect.[68] Despite a continuing trickle of reports of infant deaths due to formula-feeding mishaps, doctors appeared more secure than ever in their comparisons of breast milk to formula. In addition to being nutritionally adequate, proponents of formula declared that bottle-feeding had finally matched "convenience, conservation of time, and reduction in labor costs" sufficiently to overtake breastfeeding's traditional lead in these areas.[69] By the mid-1960s scientific formula preparation was moving out of the hospital and into specialized manufacturing sites.[70] The resultant reduction in the costs of formula meant arguments about breastfeeding's economic utility became less and less convincing. By 1967 the American Public Health Association was stating with certainty that "although breast feeding has obvious advantages . . . it turns out to be more expensive." In order to meet the recommended nutritional needs for lactation, "the mother will have to consume an extra $10.50 to $12.60 worth of food per month." Meanwhile, they stated, "a simple can of evaporated milk formula would cost only about $6.30 per month," and even a "proprietary formula," they observed, only "costs about $9.00 per month."[71]

Arguments about the cost of breastfeeding combined with waning medical faith in the long-accepted fact of its natural superiority. The "natural" suitability of breast milk for infants, in fact, came under attack with increasing frequency as researchers reported deficiencies in optimal levels of vitamins—particularly vitamin D—ascorbic acid, and iron. As one doctor put it, "when many lived naturally, in the sun and by instinctive choice of natural foods, it may have been safe to assume that mother's milk was adequate to serve as the sole food in early infancy." The reality, argued these professionals, was that man, and more specifically *woman*, was no longer living a natural life. Many reasoned that since Americans were living "the unnatural life of civilization," questioning the "perfection of Nature" seemed not only justified, but also necessary.[72] Finally, the midcentury emergence of arguments about maternal instinct and the psychological and emotional health of the mother and infant, subjects on which pediatricians had long expressed their ambivalence and discord, grew less potent, becoming almost background noise in pediatric journals. Pediatricians, inadequately trained in the psy-ences, tended to oversimplify psy-entific claims, a situation perhaps accelerated by the work of popularizers like Dr. Spock who distilled complicated psychological theories and research into simple "commonsense" wisdom. In doing so, writers like Spock not only created a canon that any mother could follow; they also informed the opinions of their peers.[73] Concerns over mother-infant attachment and the instinctual psychobiological phenomena that psy-entists identified in maternal processes became, in the hands of pediatricians, arguments for keeping insecure mothers calm and assuaging the fears of overanxious new parents. As one pediatrician put it, "if the infant gains such psychic satisfaction at the breast as we have been led by some to believe, it seems surprising that almost every infant, if he is being fed by both methods, prefers the bottle."[74] In these anecdotal arguments, it was not breastfeeding itself that assumed importance, but the physical proximity of the mother to her infant. As one author put it, "A baby held lovingly while being bottle fed has just as close a bond with his mother."[75]

Furthermore, there remained a strongly and widely expressed concern in the medical literature for the mother who showed a "repugnance" or "revulsion" toward breastfeeding. Doctors, including Spock, warned against "forcing" breastfeeding on "these kind of women," whom they identified as emotionally immature and unable to handle the commitment and intensity required of the breastfeeding mother.[76] Other physicians adopted a less judgmental perspective (though no less condescending, perhaps) and looked with pity upon the new mother and the rest of her family whom they saw struggling with the stress of bringing home a new infant. These concerns focused on the mental and emotional health of the mother, husband, and even older siblings. Some, for example, reasoned that it was cruel and harmful to subject a young family to the stress and uncertainty of a breastfeeding regimen, in which "uncertainty of adequate intake," the exclusion of the father from the early mother-infant relationship, the possibility of "failure" and the subsequent guilt that often accompanied it, and the provocation of sibling rivalry only added to the emotional and physical demands made upon the mother.[77] On the other hand, the "certainty" of bottle-feeding, in which the mother knew exactly how much food her baby was eating and the ability of all in the family to share in the feeding of the infant could reduce the stress of bringing a new baby into a family. More and more physicians certainly seemed to prefer the latter choice, enjoying a more controlled, unstressed patient and a more standardized infant.

Although couched in terms of reducing stress and anxiety, the growing lack of medical support for breastfeeding did little to help the mothers who actually wanted to breastfeed. Instead, doctors and nurses often ended up in disciplinary struggles over who exactly was to blame for poorly managing breastfeeding mothers. Physicians, including pediatricians, general practitioners, and obstetricians, as well as nurses, took turns pointing fingers. Sometimes they shouldered the blame themselves and sometimes they blamed others for not doing a proper job of preparing, assisting, or managing a breastfeeding mother. In doing so, professional struggles over disciplinary jurisdiction became intimately wrapped up in breastfeeding discourse. One need not look too hard to dig up evidence of these disciplinary squabbles over who should do what. As Benjamin Spock put it: "The obstetrician who favors breast feeding detects the itch in certain pediatricians to shift to formula, and the pediatrician who strives for breast feeding is frustrated by the obstetrician who seems eager for any excuse to wean. The occasional nursery nurse, who spends some of the most crucial minutes with the inexperienced mother . . . subtly underline [*sic*] the mother's confidence in her ability to nurse and reduce [*sic*] her to tears."[78] Not uncommonly, pediatricians laid blame for breastfeeding difficulties in their patients on the shoulders of the obstetricians, accusing them of generally lacking any interest in talking to mothers about breastfeeding.[79] As one doctor put it bluntly: "Sometimes we have difficulty in getting them [mothers] to try [breastfeeding] because they have taken a negative attitude from the obstetricians who have told them that there is not much chance of their being able to nurse. We would probably have more success if obstetricians would encourage mothers to work on their breasts prior to delivery."[80] Others adopted a more judicious outlook of the situation, as British pediatrician Harold Waller wrote: "I believe it true to say that to most obstetricians the supervision of infant feeding comes as an anti-climax to all the problems and anxieties safeguarding women through pregnancy and labour. The majority are very content to leave the care of the infant to the nursing staff."[81]

Obstetricians, for their part, were behind the increasingly widespread use of stilbestrol, the lactation suppressant synthetic hormone. Doctors utilized the drug as a way to "minimize the early increased milk tension" that they frequently observed, which could result in engorgement without proper expression. Obstetricians were also responsible for the practice of heavily sedating mothers during labor, a practice that resulted in lethargic infants often born too groggy to exercise their sucking reflex at birth. In a 1963 case study by infant behavior expert T. Berry Brazelton, for example, the obstetrician appears only once in the story of his patient's labor, delivery, and postpartum care. Described as being "determined to feed her baby at the breast," the mother explicitly stated that she wished "to have no medication during labor." Despite these wishes, her obstetrician appeared after twelve hours of active labor and "administered spinal anesthesia." Afterward, the baby, "Bill," was described as "cycling between extreme sleepiness and hyperactivity," during which he cried for periods that "lasted for 15 to 20 minutes." As the mother's efforts to feed him ended in frustration, "she became depressed and less effective in handling him." Brazleton went on to report that it was only with "constant attendance from nurses and physicians" that the mother was able to successfully breastfeed yet even then she did not seem to enjoy "him at his feedings, but seemed relieved when she could send him back to the nursery."[82] Though uncritical of the obstetrician's actions, Brazelton's case study highlighted the extent to which the obstetrician's approach to his patients impacted the mother and the job of the pediatrician in the postpartum period. Follow-up studies by others who built on this work demonstrated even more clearly that infants born under obstetric sedation "sucked at significantly lower rates and pressures, and consumed less nutrient than infants born to mothers who received no obstetric sedation."[83] Thus, as time wore on, the practices of obstetricians seemed to infringe more and more upon the health of the infants, who came under the care and authority of pediatricians after birth.

As much as pediatricians discussed obstetrical practices, however, they generally balked at interfering too much in areas that were so clearly beyond their jurisdictional authority. Obstetricians were also not above expressing annoyance over the practices of their pediatric colleagues. As one obstetrician explained to his patient, he discouraged breastfeeding in most cases because "patients called him at 2am with fissured nipples and complaints that pediatricians had told them to dry up, they thought their milk had stopped, etc."[84] Pediatricians and general practitioners often acknowledged their fields' tendency to actively discourage breastfeeding and to instead rely on bottle-feedings for their manageability, convenience, and availability. As one savvy mother put it, "I cannot fathom why pediatricians are so quick to say 'dry up and use [C]arnation.' Either they own stock in the company or they are afraid of something that they can't see measured in a bottle in exact cc's."[85]

Pediatricians and general practitioners were often insecure about infant weight gains and many generally found the control they exercised over formula feeding reassuring.[86] If they saw themselves at fault in the low rates of breastfeeding, however, it was due to how they managed their patients while still in the hospital. In the realm of hospital care of the mother and newborn, doctors frequently pointed out that breastfeeding lay in the hands of the staff nurses. Dr. Harold Waller, a staunch proponent and keen observer of lactation and all its complexities based at the British Hospital for Mothers and Babies in Woolwich (London), was perhaps more sympathetic than most in his assessment of the nurses' roles. He reported that he could usually tell when the nursing staff had been "overworked" due to a "rush of admissions" by observing that his patients' charts revealed an unusually high reliance on lactation suppressants and "there may be two or three mothers with injury to the nipples."[87] All of which suggested that the nurses had not had enough time to adequately instruct mothers on appropriate breastfeeding techniques.

These disciplinary squabbles of course did little to improve the overall experience of breastfeeding mothers who, along with their infants, found themselves at the center of a professional quandary that went beyond the question, Is breast best? Regardless of whether physicians or nurses believed breastfeeding was better than bottle-feeding, a significant number of mothers wanted to breastfeed their babies. The fundamental question health care providers were at pains to address was, How do you manage the mother who wants to breastfeed within a model of scientific medicine? As the overall rates of breastfeeding mothers continued to fall through the 1960s, it became increasingly unlikely that physicians would be prepared with the education, experience, and resources necessary to help a breastfeeding mother. Mothers who chose breastfeeding in the midst of the bottle-feeding age faced a particularly uphill battle.

Confronting the New Breastfeeding Mother

At the close of the 1940s, two-thirds of mothers were still breastfeeding at the time of hospital discharge. A decade later that percentage had dropped to less than half. What statistics like these miss, however, is the emergence of a new kind of breastfeeding mother in the midst of this larger decline. Emboldened by the growing science and ideology of natural motherhood, women with access to education through childbirth and breastfeeding support organizations and ample resources of time, money, and social capital began to resist the inertia of hospitalized childbirth and postpartum routines. As historians such as Judith Walzer Leavitt and Jacqueline H. Wolf have shown, the vast majority of women by the 1950s were giving birth in hospitals, their pain managed by an array of interventions by medications, doctors, and nurses.[88] After birth, nurses typically whisked away newborns to keep them under close watch in a nursery for the first twelve hours of life or more. Only after a lengthy period of medical monitoring were infants brought back to their mothers to get acquainted and to eat. For first-time mothers especially, the circumstances under which they entered motherhood deprived them at every stage of their autonomy and authority. By the time they left for home, most would have had little opportunity to build up confidence in their abilities to take care of their frighteningly dependent charges. Of those who were still breastfeeding at the time of hospital discharge, the majority would have learned not to try to do so exclusively. It was a common practice to combine supplemental feedings after an infant finished off a breast for fear that the infant might not get enough to eat in the first few weeks of life. It was just as common for medical sources throughout these decades to suggest that a daily "relief" bottle could be a sensible, if not a sanity-saving, measure for any new mother.[89] In the face of this institutional inertia, however, the emergent ideology of natural motherhood emboldened small numbers of women to successfully persevere. Physicians took notice and occasionally commented on the appearance of this new kind of mother, and for those who continued to support breastfeeding, she was a welcome sign of change.

Many health care providers simply did not know what to do with this "new" kind of patient. Some reached out for assistance from the network of childbirth education organizations that had grown up around the country. In 1957, for example, the chairman of St. Margaret's Hospital in Boston, Massachusetts, wrote to Justine Kelliher at the Boston Association of Childbirth Education, noting, "In the past few months we . . . have noticed an increase in demand for natural childbirth, rooming-in, etc." He asked if they might send someone to speak to their staff and patients on the issue.[90] One physician, Dr. Howard J. Morrison (a breastfeeding supporter), was so struck by his encounters with this emerging group that he waxed poetic about his observations of four natural childbirths in the early 1950s, "These mothers are amazing," he wrote, before remarking on their peculiar yet admirable desire to "hold or have their baby close enough to touch while still in the delivery room." Compared to the majority of births he had attended in which mothers had "delivered under heavy analgesia or anesthesia," the mothers in these natural births astonished him with their "euphoric" moods immediately after delivery. A state, he noted, which persisted even "while expelling the placenta or having the episiotomy repaired."[91] Morrison further observed that each of these mothers "successfully breast-fed their infant," a phenomenon that prompted him to reorient how he thought about the relationships between pregnancy, childbirth, and infant feeding. "Natural childbirth, breast feeding, and a 'self-demand' schedule of feeding appear to be intimately related to each other," he mused.[92]

In a less expressive, though no less enthusiastic, discussion of a growing interest in "naturalism," childbirth, and breastfeeding practices, Dr. Edith B. Jackson recorded similar observations through statistical analysis. Utilizing patient samples from both the rooming-in program and the more typical "nursery mothers," Jackson and her team at Yale found "that the percentage of all mothers who breast feed in the hospital decreases with the increasing difficulty of the delivery."[93] Jackson assigned delivery scores based on the kind and degree of interventions by medical staff on a woman's labor. The lowest score, a 3, indicated that a woman had received no anesthesia, had experienced no more than twelve hours of labor, and had delivered without forceps. Higher scores went to women who had used anesthesia to varying degrees, had labored for more than one day, or whose deliveries involved the use of forceps or culminated in cesarean section. Jackson found that women who experienced non-traumatic labors, with scores of 3 and 4, made up the bulk of the breastfeeding mothers in both the rooming-in and nursery groups. This kind of data contributed further to evidence that breastfeeding seemed to follow the same pattern as the labor and delivery, with physically and psychologically difficult deliveries resulting in fewer breastfed babies.[94] Of even greater consequence was Jackson's finding that of those mothers who chose breastfeeding, rooming-in mothers on average breastfed for an additional month compared to nursery mothers. The most provocative of her findings, however, suggested an important connection between the answer women gave when asked "why breastfeeding?" and their overall success and duration of breastfeeding. Mothers' answers were recorded and then grouped into six categories, including "better for baby, better for mother, convenience, influence of relatives or doctor," and "previous experience." But it was the sixth category,.which Jackson dubbed "naturalism," that prompted her to observe: "There was a definite indication that the group of mothers who wanted to nurse because it was 'the natural way' had the longest mean duration."[95] She went on to elaborate that "the wish to breast feed because it is 'the natural way' may in itself be related to longer duration of breast feeding. Quite possibly this wish may intensify a mother's motivation for rooming-in as also 'natural' thus influencing the rooming-in group in favor of longer duration of breastfeeding."[96]

Despite observations by those like Morrison and Jackson and others throughout this period, the vast majority of doctors were not ready to simply give control over the birth and postpartum processes to mothers or to "nature." Even many of those who supported breastfeeding in theory struggled with how to medically manage a breastfeeding mother and infant within this rising framework of "the natural." Maternal and child health experts argued over what seemed like an endless array of variations in the prescribed nursing routine: Was it best for an infant to drain one breast per feeding? Or was it better to feed from both breasts every time? How long should feedings last? One minute? Five minutes? How frequently should the infant be put to the breast? Should supplements of water or formula be offered after feedings? In cases of engorgement should women manually express milk? Should they use a breast pump? Physicians and nurses alike bickered over the details as they attempted to iron out a standard, scientific regimen for breastfeeding management that also allowed for the kind of individual variation they knew existed.

Mothers encountered a wide range of breastfeeding advice, practices, and policies before, during, and after their admission to the hospital maternity ward. Doctors sometimes performed breast examinations in order to evaluate the mother's nipples prior to the onset of labor or shortly afterward. Mothers could be diagnosed as having any number of "abnormal" types of nipples that could play a role in an unsuccessful breastfeeding attempt, including everything from the "mushroom" to the "mulberry."[97] Doctors and nurses supported elaborate breast-care routines both during pregnancy and afterward, sometimes as complex and time-consuming as mixing up a bottle of formula. Unlike mixing formula, however, mothers generally received very little care and attention from their health care providers when it came to actually breastfeeding. With detailed instructions and their physician's prescription, mixing formula was a science—a set of practices that could be taught and implemented using familiar kitchen procedures. Breastfeeding, on the other hand, was understood as an art, one that required the acquisition of tactile knowledge and the cultivation of bodily intuition. Unlike formula feeding, which changed only in concentration as the baby grew, breastfeeding was a dynamic and largely invisible process, an emotional and physiological experience in constant flux as the infant grew and the mother's body responded. Physicians who encountered breastfeeding mothers throughout this period struggled to find ways to fit the practice into a controlled scientific regimen, and generally failed in doing so.

Despite evidence that hospital and medical routines themselves were the culprits behind much of the lactation failure that mothers experienced, physicians and nurses cited nipple difficulties and other breast complaints as the most common causes of breastfeeding failure in their patients. As a result, they developed a wide variety of protocols to address these kinds of problems. It was not uncommon, for example, for physicians to advise a woman with "abnormal nipples" that she would not be able to nurse because her infant would not be able to suck adequately and she would likely experience a lot of pain. The dreaded "inverted nipples" were the ones most frequently discussed. Some doctors tried to convince mothers with this anatomical feature to forgo any attempt to breastfeed, while others suggested the use of a nipple shield during pregnancy or even the manual stretching of the nipple by the woman in the months prior to delivery.[98] Even mothers with "normal" nipples were frequently advised to participate in pre-labor breast preparations including the use of various moisturizers such as cocoa butter, lanolin, or even cold cream.[99] Alternatively, toughening procedures such as the use of rubbing alcohol and a "stiff bristle brush" to clean the nipples daily in the weeks prior to delivery were also advised in an attempt to ward off injuries. Some physicians and nurses advocated methods to break the infant's suction at the breast that saved the mother's nipples from undue trauma at the end of a feeding session. As one physician advised, "Hold the baby's nose to make him stop nursing rather than . . . [pulling] him off the nipple."[100]

With a few exceptions, rarely did physicians or nurses mention anything about how to actually get an infant to "latch on" properly. Now widely understood to be a key component of successful breastfeeding, getting a "good latch" is something that consumes the mother and her postpartum nurse and/or lactation consultant in most hospitals today.[101] While a handful of practitioners advocated preparations that would facilitate the infant's latch, such as teaching mothers to manually express colostrum prior to delivery and supporting natural childbirth practices so that infants would be born alert and ready to suck immediately or soon after birth, the majority of obstetricians did not.[102] Instead, for most of the mid- to late twentieth century, physicians and nurses alike focused on time at the breast, believing that, along with nipple preparation and shape, the time the infant spent sucking at the breast was something that required careful management. The only problem was that there were nearly as many suggested nursing regimens as there were experts, a fact that no doubt contributed to a sense of confusion and hopelessness in many breastfeeding mothers.

Despite the almost infinite variations in nursing advice, however, tracing the practices and advice of health care providers throughout the postwar decades reveals a clear set of continuities in how physicians approached breastfeeding management in this period. As early as 1950, physicians like Frank Richardson were calling out hospital policies which set an "*unnatural* limitation to twenty minutes at a nursing,"[103] while another journal in the same year advanced the idea that "initial feedings require about 5 minutes at each breast," but once the "baby has become stronger, about 15 minutes are required to satisfy the needs of the infant."[104] Others maintained far more complicated routines that combined scheduled three- and four-hour feedings with limited nursing times, such as the case with Howard J. Morrison's patients at the United States Naval Dispensary in Washington, DC:

> [Infants] were put to breast at the first regular feeding time after they were 12 hours old and allowed to nurse for 5 minutes at each breast 3 times a day until the third day when a 4-hour nursing schedule was instituted day and night, the infant being allowed to nurse one breast at a feeding as long as it desired. Five percent glucose water was offered every 4 hours until lactation was established, then boiled water was offered after breast feedings. If an infant took more than an ounce of water after each feeding, a complementary formula was offered.[105]

Practices of limiting sucking time to as little as one minute,[106] spacing nursings out over periods of three or four hours, and supplementing feeding with prelacteal and post-breastfeeding bottles consisting of everything from boiled water to glucose solutions to cow's milk–based formulas all persisted well into the 1970s.[107] Additionally, misunderstandings about the bacteriological safety of breast milk led many to institute rigorous hygiene regimens before every nursing session. Women underwent everything from bathing with soap and water solutions before and after every breastfeeding to the application of antiseptics. It was common for health care providers to advise women to maintain hygiene regimens that singled out the breasts for special attention, such as washing "the breasts before washing any other part of the body."[108]

This kind of medical advice, gleaned from clinical experience (at best) and assumptions based on physicians' experiences with formula feeding and contagion, is likely what Dr. Clement Smith referred to as "the absence of modern factual knowledge" about breastfeeding. It was also what Dr. Edith B. Jackson described when she wrote, "Nurses and doctors seemed to have no time and little understanding for helping [mothers] and their infants adjust to the process of breast feeding."[109] Ironically, physicians and nurses themselves were in the best position to see the negative consequences of the implementation of this information. In a 1952 article, Dr. Emily Bacon, emeritus professor of pediatrics at the Woman's Medical College of Pennsylvania; chief of pediatrics at the Mary J. Drexel Children's Hospital, and chief of new-born service at the Woman's Hospital in Philadelphia, relayed the following disturbing scene:

> Instead of being fed when he is hungry, [the newborn] often has to succumb to a man-made rigid rule. It is not unusual to make rounds in a newborn nursery an hour before the scheduled four hour feeding, and find a baby crying, kicking off his clothes, rubbing his heels or knees nearly raw, chewing his fists, scratching his face and otherwise vigorously "rooting" for food. By the time he is given his feeding, his stomach is full of air swallowed while crying, and he is too exhausted to nurse well either at the breast or bottle. The baby should be fed when he needs food . . . he then nurses peacefully, the mother is happy and at ease, and consequently her breasts function better.[110]

These experiences led her to conclude: "So far as his feeding is concerned, it is a disadvantage for a healthy baby to be born in a hospital."[111] Despite the cacophony of untested breastfeeding regimens that stressed toughening the nipple, limiting nursing times, and relying upon additional (usually formula) feedings, dissenting voices could be found scattered throughout the literature. Niles and Michael Newton, for example, continued to publish on the requirements for breastfeeding success based on actual clinical and experimental data. In 1961, Michael and two of his colleagues published a clinical study of the relationship between number of feedings at the breast and the overall milk supply. They showed that "mothers who wish to increase their milk supply should be urged to feed their infants two or three extra times each day" than doctors typically prescribed.[112] The Newtons also suggested that medicine's enthusiasm over cleaning nipples before and after feedings was more injurious than not and instead recommended the use of lanolin on sore nipples and limited hygiene regimens.[113] They argued further, based on existing medical data, that the common practice of ending breastfeeding at the onset of breast problems such as sore nipples, abscess, and mastitis was an unproven and if anything largely harmful practice. Rather, they argued, treatment with antibiotics, support, and continued nursing seemed to clear up common breastfeeding problems faster while keeping breast milk plentiful.[114]

The 1960s

By the late 1960s, even while the percentage of breastfeeding mothers continued to shrink, there remained those in the medical system who continued to express enthusiasm, if not nostalgia, about working with breastfed infants. "I am delighted when mothers tell me they plan to nurse their infants," wrote pediatrician Dr. Lee Forrest Hill from Des Moines, Iowa. "For those mothers who are on the fence, I do not hesitate to extol the virtues of nursing and to express the opinion that human milk is superior to all other milks," he added.[115] Gregory J. White, a general practitioner in Oak Park, Illinois, throughout the 1950s, 1960s, and well beyond, was another longtime supporter and advocate of maternal health reform. As the husband of Mary White (founding La Leche League member), Gregory White saw firsthand through his wife's experiences, and those of her friends, just how difficult it was for women who wanted to breastfeed. He became known throughout his community for his support of natural childbirth, breastfeeding, and even performed home deliveries well into the second half of the twentieth century.[116]

Dr. Edith B. Jackson moved to Denver in 1960, where she quickly went to work applying what she had learned at the Yale Rooming-In program to the newborn nursery there. "We had been and were continuously interested in teaching our medical and nursing staff about breast feeding," she wrote, and was glad by that time that she could "refer appropriate breast-feeding patients for help to La Leche League members," in the area.[117] At Colorado General Hospital, "Edie" (as she was known) designed a new rooming-in unit and implemented many of the same principles of mother and newborn care that she had developed while at Yale.[118] Her support for breastfeeding and mothers went beyond the confines of the hospital. Jackson was a longtime supporter of women's health networks. "I participated in some of the earliest childbirth education groups in New England," she wrote, "and was wholly supportive of this type of group." Jackson, despite her own position as a physician, was explicit in her belief that medicine could only do so much for breastfeeding mothers. "I have been thoroughly sympathetic with the organizations of mothers to help each other," she wrote, particularly "when they fail to find the professional help which they need."[119] Jackson served on La Leche League's Medical Advisory Board (MAB) in the 1960s, along with over a dozen other breastfeeding advocates from the medical community, including obstetricians, pediatricians, psychiatrists, and nurses from across the United States. Among her peers on the MAB was Dr. Robert S. Mendelsohn, a Chicago pediatrician who helped run the city's Head Start program and maintained outspoken views against hospital childbirth, the pathologization of pregnancy, and the lack of medical enthusiasm for breastfeeding.[120] Over time, Mendelsohn became more and more critical of medicine and even went so far as to encourage people to "listen to 'old wives tales' and grandmothers, who frequently know more about childrearing than doctors do" and believed home births were safer than those in hospitals.[121]

That women could, albeit with great difficulty, find doctors who would support them in their desire to breastfeed is evidenced in the writings of mothers themselves from this period. In 1960, one mother wrote to Dr. Spock to share her breastfeeding story along with some suggestions for how he might improve his published advice on the subject. "I am apparently one of the 'infrequent cases,'" she wrote, somewhat pointedly, who had suffered from "very painful" engorgement. Luckily, she explained, she had a medical support team in place to help her. In addition to having a rooming-in setup for her birth, she delivered her baby in a maternity ward in a Catholic hospital "run by a little wisp of an Irish nun," who made sure that even at night the baby was brought to her "any time he cried." She credited her eventual success to determination, "pumping twice a day," the "use of a nipple shield [which] softened the breast and allowed the nipple to be drawn out," and a medical team, including an obstetrician, a pediatrician, and a head nurse, who all made sure she received the support and encouragement that she needed.[122]

Observations, ideas, and practices of medical practitioners like those who attended the mother described above, however, remained on the periphery of mainstream care and did not deeply penetrate the medical profession. Often doctors and nurses believed that only a certain kind of woman, an exceptional kind, would have the perseverance to undergo the self-sacrifice they believed it took to successfully breastfeed. The vast majority of women who wanted to breastfeed, in fact, continued to confront what could be an overwhelming lack of information and support when they attempted breastfeeding in the 1960s, just as those in the 1950s had. Throughout these decades, most physicians generally exhibited what Dr. Lee Forest Hill described as "a deplorable lack of interest in the mode of feeding."[123] Certainly, as doctors themselves observed, the overall interest in breastfeeding among physicians dwindled over the course of the 1940s, 1950s, and 1960s and these circumstances conspired to make usable clinical knowledge about breastfeeding scarce, at best. By the late 1960s, medical interest in breastfeeding seemed to pass almost entirely out of the hands of physicians altogether, as lay groups such as La Leche League and, increasingly, nurses, emerged as lactation experts, authors, and advocates. Doctors who supported breastfeeding in theory but struggled to make measurable gains in practice instead came to rely on lay groups like the League (formed in 1956) to manage their breastfeeding mothers for them and even occasionally expressed relief that there was someplace to which they could refer women when they encountered breastfeeding difficulties.[124]

Despite a persistent network of mothers and health care providers dedicated to breastfeeding, hospital management of labor, delivery, and the postpartum period combined with the ongoing expansion of commercial baby formulas and physician apathy to push breastfeeding further into the margins of medical practice.[125] By 1967, breastfeeding rates were so low that pediatrician Lee Forest Hill could write about it as though it was of a bygone era. "To those of us who recall the days when human milk was esteemed above all other foods in infant nutrition, it is distressing indeed to observe how lightly it is regarded today," he lamented.[126] Well into the 1970s it remained standard practice in many places for mothers to be separated from their infants, sometimes for as many as twelve hours immediately after birth. Similarly, even while a movement toward including fathers and other family members in the labor and delivery room had been building since the 1940s, it was not uncommon for a mother to continue to labor in isolation from friends and family into the 1970s.[127] Just as they had for decades, these kinds of hospital management policies kept the proportion of women who initiated breastfeeding low. Doctors in this era did not find a way to successfully and uniformly manage breastfeeding, and those who remained breastfeeding supporters often acknowledged that the practice seemed in and of itself to be ideologically and practically incommensurable with the institutions and practices of modern medicine, as it had come to exist. Those in the medical system who understood the complexity of breastfeeding and believed in its importance, then, also had to agree to relinquish a great deal of control over their patients, turning them over instead to the growing networks of mothers and grandmothers who had been consuming the science of breastfeeding for decades and who had the additional benefit of having breastfed their own infants.

By the end of the 1960s, it had become clear to many in the medical community that the reigning ideology of the era, scientific motherhood, had failed to successfully incorporate breastfeeding. While there were important medical advocates scattered across the mid-twentieth century, few of them had success integrating breastfeeding into the medical mainstream. Attempts at making breastfeeding scientific generally failed due to lingering inadequacies in research, disputes over what kinds of science counted, a growing lack of education and interest in lactation among doctors, the continuation of modern birthing practices that interrupted complex physiological processes, and a widespread belief among medical professionals that formula feeding matched breast milk in terms of its safety, efficacy, cost, and convenience. The legitimacy of the science of emotion and human behavior remained in question when it came to maternal and infant health care, and the inability of female researchers to secure and maintain positions of security and authority in their field was a likely contributor to the ongoing marginalization of relevant work by people like Jackson and Newton. Natural motherhood ideology and an interest in breastfeeding, however, continued to live on throughout this period. Mothers themselves played a critically important role throughout the 1940s, 1950s, and 1960s in keeping breastfeeding knowledge and practice alive through participation in women's health education networks, popular magazines, and the experience and support they offered to friends and family. Those who wanted to breastfeed and were successful in doing so were those who learned which doctors they could go to for help, how to work within (or sometimes, outside of) the hospital system to their best advantage, and how to deal with their nurses who could be either friend or foe. In the chapters that follow, I explore the culture surrounding breastfeeding mothers and nurses to better understand how natural motherhood ideology shaped the infant feeding landscape even as bottle-feeding continued to grow more and more ubiquitous in American culture.

CHAPTER THREE

"Motherhood Raised to the *n*th Degree"

Breastfeeding in the Postwar Years

In her essay "Breast Fed Is Best Fed," Eleanor Lake declared unequivocally and with a touch of humor that "breast feeding, once as outmoded as the bustle, is triumphantly on its way back." This is not a headline pulled from a 1990s women's magazine, nor even one from the 1970s. Lake's 1950 piece in *Reader's Digest* explored what she characterized as common misconceptions and beliefs about breastfeeding, particularly the falsehoods that breastfeeding would "tie" women down and ruin their "figures." She also boasted that some hospitals where breastfeeding rates had fallen as low as 20 percent in the early 1940s could claim just ten years later that they were sending "three quarters of all their babies home at the breast." Furthermore, she wrote, these trends were being pushed in some of the nation's "leading hospitals" and by some of the country's most "high-powered" doctors, including Benjamin Spock and C. Anderson Aldrich, among others. Lake enthusiastically explained all of the ways in which breastfeeding was best for the infant, including better weight gain, fewer allergies, stronger immunity, and healthier skin and digestive tracts. She even went so far as to highlight the ability of breastfeeding to promote a strong emotional connection between the infant and mother. Lake's article is even more noteworthy, however, for its lengthy discussion of all of the benefits for the breastfeeding *mother*, including regaining her pre-pregnancy body more quickly, balancing out postpartum hormones with fewer bumps along the way, and perhaps most importantly, contributing to an overall improvement in her emotional and mental well-being. As one mother put it, breastfeeding made her feel that she was "handing [her baby] something precious and priceless, some special kind of love that human mothers have given their babies for thousands of years." It was, in her words, "the most peaceful, joyful feeling" she had ever experienced. Lake, however, perhaps summed up these sentiments best when she wrote, "Holding her infant close, giving him a direct, nourishing love, a new mother feels an almost mystic sense of unity with her baby. Nursing is motherhood raised to the *n*th degree."[1]

This chapter explores the culture of motherhood surrounding breastfeeding in the postwar era, 1945 through 1963, focusing primarily on the ways women encountered and made sense of natural motherhood and breastfeeding in their own lives. Though many more mothers turned to formula feeding as the postwar era ticked along, there continued to be a place for discussion about breastfeeding, one that increasingly reflected its connection to the era's nascent natural motherhood ideology. Although they were often relegated to the margins of society, the women who advocated for and pursued breastfeeding during this period not only helped keep knowledge and interest in breastfeeding alive; they provided an interested and critically engaged audience for popular advice authors, researchers, and clinicians. Talking about and exchanging information on breastfeeding gave mothers the opportunity to negotiate postwar ideas about female sexuality and marital relationships in addition to maternal roles, as breasts turned into battlefields over domestic authority. In their efforts to carry breastfeeding into a new era, these women charted an alternative course for modern motherhood, one that embraced a biology of sentiment and meaning.

Maternal Expectations in the Postwar Years

Throughout the postwar years and in tandem with the rise of natural motherhood ideology, mothers slowly came to expect that breastfeeding offered a connection to an instinctual maternal identity that could be tapped into, infusing the tasks of infant rearing with personal fulfillment and a sense of purpose.[2] In attempting to actually breastfeed their children during these years, however, mothers confronted a tangled web of conflicting advice.[3] The irony in this story is that feeding babies "naturally" was of course not easy, particularly in the postwar years. Women who tried to breastfeed often quit out of frustration, worry, or because a trusted medical adviser told them to. Mothers also struggled to balance their own expectations of what breastfeeding and motherhood should be with the expectations of society, their families, and their own babies.

Even as the ideology of scientific motherhood encouraged women to seek out the most modern, scientific, and technologically advanced means of tackling their domestic and child-rearing duties, the alternative ideology of natural motherhood pulled them in the opposing direction. The work of historian Alexandra Rutherford highlights the fact that mothers in this period confronted a complex and slippery set of ideals when it came to the integration of technology into their child-rearing practices. In a discussion of psychologist B. F. Skinner's "Baby Tender," Rutherford explains how the radical behaviorist believed that human consciousness lay entirely outside the realm of scientific inquiry. For Skinner the extrapolation of this belief led to his argument that sensations and emotions were not the cause of human behaviors, but rather the effects. Therefore, control over the physical events within the body, he argued, resided in the material world not in the realm of the psyche or of emotions. Given this orientation, Skinner built a contraption he called the "Baby Tender" for his infant daughter Deborah. The Baby Tender was a climate-controlled crib meant to provide the perfect physical environment for babies when they weren't being diapered, fed, or bathed.[4] Unfortunately for Skinner, popular response to his creation (which was never widely manufactured) was incredibly negative. Skinner later defended himself saying: "It is not an experimental apparatus. It is not soundproof; Deborah was shielded from loud noises. But we could hear her at all times. It is not germproof. . . . It is no more mechanical than a standard crib, and there was nothing mechanical about the care we gave our child."[5]

Regardless of the mechanical details of Skinner's Baby Tender, Rutherford's exploration into the public reception of Skinner and his work suggests that his creation touched a collective nerve. While the technology of the Baby Tender fit the era's interest in labor-saving home-care devices and responded to the ideology of scientific motherhood, it did not fit the rise of popular psychological ideas about the developmental importance of early infancy, maternal instinct, or American's growing discomfort with mechanistic models of humanity that dismissed the less tangible components of existence. In short, it was a technology that pushed infant care into the realm of the unnatural, an argument that should not have carried much weight in the 1940s. No doubt Skinner's daughter was raised with no less physical contact than most American babies born in 1944, the tail end of the era of timed feedings, minimal coddling, and strict rules for managing everything from diapering to crying. However, the very idea of the Baby Tender, in all its mechanical physicality, made the physical and emotional distance between baby and mother all the more obvious and disquieting. Furthermore, as Rutherford suggests, Skinner's baby box hit on rising anxieties about the proper role of the mother. After all, if a box raised one's baby, what exactly was the job of the mother?[6] Historians have pointed out that during this period the proliferation of modern household technologies generated a widespread popular conception of the American housewife as idle and carefree, save for her increasingly important role as a mother.[7] Skinner's Baby Tender threatened that final stronghold of female household responsibility and seemed to question the historically unshakeable role of women as child rearers.

Skinner's invention, and its failure, brings into relief the tension in American culture at the time over the changes that were taking place in the opportunities and experiences available for women in everyday life. The public's reaction to Skinner and his work highlights Americans' growing concern over the psychological and personality development of infants. More importantly, however, I suggest that the public's reaction to Skinner's baby box hints at deeper social anxieties over the changing roles of women. The primary cause of the worry surrounding the invention was that it suggested mothering was something that anyone, perhaps even any*thing*, could do. Despite its ultimate failure as a consumer technology, the controversial baby box prompted people to consider what exactly women's roles would be in a world in which infant care and mothers had been decoupled.[8] The negative reaction to the Baby Tender offers a glimpse into the subliminal anxieties of mid-twentieth-century American culture, of which motherhood was exceedingly important. It was precisely this vein of anxiety that the ideology of natural motherhood tapped into, giving space and voice to a growing discourse on the place of breastfeeding in American households.

Breastfeeding Guilt

It was highly probable that mothers in the postwar years would have come across discussions of the psychological, emotional, and physical benefits offered by breastfeeding while flipping through a magazine, pamphlet, or health manual.[9] Wherever discussions on breastfeeding surfaced, advisers made it clear that breastfeeding was best, though they were usually quick to point out that most mothers were not "ideal."[10] Tracing the discussion surrounding guilt and breastfeeding "failure" in women's popular magazines and mothering advice literature helps reveal what breastfeeding meant to mothers in this period. Rather than happily embracing bottle-feeding, this evidence suggests that many mothers struggled to breastfeed in an age rife with inadequate knowledge and medical support. That they made this choice suggests their hope in the promise that breastfeeding could offer them something more than a cheap and convenient food source for their babies.

In a 1956 letter to Dr. Benjamin Spock, *Ladies' Home Journal* editors Bruce and Beatrice Gould reflected that "the Journal [has] for a long time been beating the drums for breast feeding . . . because we thought all things considered, it is better to breast feed than to bottle feed."[11] The message behind this popular advice was clear, a bottle-feeding mother "is not giving the child herself; she is faithfully, efficiently providing the child with a bottle—an object."[12] Spock, for his part, expressed concerns over the consequences of touting the benefits of breastfeeding too enthusiastically. In a 1955 letter to Edith Jackson, he asked for her advice on an article he had written for *Ladies' Home Journal.* The piece had been "designed to keep young parents who can't have natural childbirth, rooming-in, and breast-feeding from feeling too badly." "Does it strike a fair balance, and is it accurate enough?" he inquired.[13]

Readers of these women's magazines, too, chimed in with their own experiences and advice. One mother, writing in *Ladies' Home Journal* in 1952, urged those who resorted to bottle-feeding to make sure they did not succumb to the temptation to "prop" bottles or leave babies in the care of other family members during feeding times. She wrote, "Don't get into the habit of letting someone else feed the baby or of propping up his bottle so that he can feed himself." "If you do," she warned, "you will be missing some of the very best moments of being a mother."[14] Another mother wrote to Dr. Spock to ask him to make a case in his *Ladies' Home Journal* column for sustained skin-to-skin contact between a mother and infant, particularly if the infant was being bottle-fed. She suggested that bottle-feeding and mental illness rates were linked and that, in lieu of actually breastfeeding, mothers and fathers should employ skin-to-skin contact during their feeding sessions. "I sincerely feel," she wrote, "that this 'cheek to bare arm technique of feeding' should be more fully exploited with the hope that more natural security develops in our children."[15] Her letter went on to imply that some mothers seemed to just *naturally* know what an infant needed, and some did not. This concern applied, of course, only to bottle-feeding families who many believed needed to take special care not to deprive their children of the experience of physical touch.

Despite the public discourse and advice on the natural benefits of breastfeeding, the vast majority of physicians, pediatricians or otherwise, did little to actively change the course of breastfeeding among their patients. Some stated outright a preference for the manageability of a bottle-feeding mother, and physicians throughout this period generally lacked a practical understanding of human lactation.[16] It was not uncommon for doctors to privately assess the validity of a mother's claim that she really wanted to nurse. As one obstetrician explained in 1950, when he determined that a mother didn't truly want to breastfeed, he simply told her that she was "unable to, because of inferior milk or some such pretext." In doing so, he believed that "if the baby were to get sick or die, she might have fewer guilt feelings."[17] Doctors also feared that if mothers who did not actually want to breastfeed felt forced into it, they might do more harm than good. Such an "emotionally disturbed mother," reflected Dr. Howard Morrison, would be unlikely to provide "security or even gratification" for her infant and was better off bottle-feeding.[18] In addition to the doubts of physicians and nurses, the maternity ward itself was set up to maximize efficiency and hygiene and it was bottle-feeding, not breastfeeding, that fit the hospital's routine. Critics blamed everything from overworked nurses, to hospital scheduling, to undereducated and unfeeling doctors for hospitals' failures in promoting breastfeeding.[19] It was common practice for nurses to give infants bottles shortly after birth, often without the mother's knowledge or consent, and "dry up" pills and injections were often administered as a standard of care.[20]

Not all physicians acted to undermine their patients' desires to breastfeed, of course. As demonstrated in the previous chapter, mothers could and did find help and care from sympathetic practitioners. La Leche League served as one site where local mothers swapped insights, tips, and references for doctors and hospitals. The practice of Dr. Gregory White was no doubt boosted by his wife Mary White's role in breastfeeding education and support in their community, and evidence suggests that other doctors who were willing to embrace women's groups like this also became important nodes of access to the kind of childbirth and postpartum care mothers wanted. The Boston Association for Childbirth Education served as a networking organization in this way, linking mothers to local physicians who could serve their needs. Such information spread not only through informal conversations at group meetings, but also through formalized invitations to speak and to attend classes as guest educators. A 1963 letter from the group Nursing Mothers Council to a Dr. Cochrane at the Boston Lying-In Hospital reflects this phenomenon:

> We have heard much about you through Miss . . . and several mothers who have delivered at BLI. This association is always pleased to know of doctors who are enthusiastic about nursing. One of my neighbors, who was not successful at nursing her first two babies, has been happily nursing her third for several months. I feel she would not have had this satisfaction had you not given her encouragement in the hospital. . . . We would be pleased if you would speak to us this year.[21]

While only some women had the knowledge and connections to access these networks, it is likely that there were others like White and Cochrane who did everything they could to help their patients breastfeed. For many women throughout this period, however, getting out of the hospital was the first of many hurdles. Mothers who made it out of the hospital with their milk and determination intact continued to face fears of failure. As one expert explained, "The mother who attempts to nurse her infant, but for some psychic reason does not succeed, suffers from a frustration that comes from denial of the nursing instinct." Advice articles reminded mothers that if they *really* wanted to nurse, they could do so. Not surprisingly, then, inherent maternal flaws could be revealed through breastfeeding failure. Failing to successfully breastfeed might be judged by experts, and by the mother herself, as the result of some buried maladjustment or a secret desire to be unbound from her infant in order to "legitimately resume her social life."[22]

The judgments that mothers faced only added to a mounting anxiety over breastfeeding. One author wrote, "Failure to nurse a baby is most often blamed on an insufficient quantity of milk," a condition caused when the mother "mismanaged" the feeding in some way. The least sympathetic mother in these situations was often chastised for not taking her new role seriously enough. *Parents' Magazine* warned its readers about the young mother who made the mistake of returning to her regular household duties immediately after returning from the hospital. Though her milk had seemed "ample at first, it soon decreased," which prompted her to ask her doctor for "a formula for supplementary feeding." She entertained visitors and when "company came she did not always find it convenient to nurse the baby; so she placed him in his crib with the bottle on a pillow beside him." Coupled with her habit of nursing him at "odd" and "irregular intervals," the supplemental feedings contributed to diminished milk production. By the time the mother realized something was wrong, "the baby was having diarrhea from near-starvation." Saved from the brink of death by a physician's "strengthened formula," the baby recovered and began to "gain at an amazing rate." The moral of the story was clear: "This young woman could have nursed her baby, if she had been wiser."[23]

Mothers could often be their own worst critics. Given the implications of breastfeeding failure, women often suffered through breastfeeding problems and blamed themselves if and when things went poorly. Despite enduring horribly sore nipples that eventually forced her to stop nursing, one mother argued that she "firmly believe[d] that nearly all mothers can breastfeed their babies *if they want to badly enough*."[24] Of course the most ironic consequence of the link drawn between a woman's psychological fitness for motherhood and her breastfeeding success was that mothers had one more thing to worry about: becoming too upset or overworked or tired, or even feeling resentment toward the wailing infant at her breast, all were warning signs of maternal failure. Conveying the anguish that such an experience could have for a young mother, another woman explained to the audience of *Parents' Magazine* that she "had cried with disappointment when [she] was forced to give it up at six weeks because of sore breasts and scanty milk."[25]

Despite the lack of support for mothers, the vast majority of advice for breastfeeding success suggested that it was women who ultimately bore the brunt of the responsibility when it came to making nursing work. Published advice told mothers to maintain a proper diet, keep social obligations to a minimum, and to make sure to get plenty of rest while breastfeeding. Much of the published advice did little, however, to make breastfeeding sound like anything other than a self-sacrificial obligation that any mother should want to do. As one author explained, "If the mother could only realize that six or seven months of secluded living during her average lifetime of 65 years is comparable to giving up one drop of water from a glassful, she would willingly want to nurse her baby."[26] The phrases "don't worry" and "get plenty of rest" were doled out frequently and in earnest as ways women could avoid the crushing failure of finding themselves with "insufficient milk." Making the care of one's self a priority in the weeks and months after the arrival of a newborn, however, was unlikely. In addition to the burdens of taking care of a new baby, mothers found their household duties waiting for them when they returned from the hospital. One author suggested that "until the milk is well established (usually after two or three months) it is best to do as little housework as possible . . . and to get at least a two hour nap in addition to your 8 or 9 hour sleep at night."[27] Combining breastfeeding with the demands of the modern housewife was exhausting, the authors acknowledged, but not impossible. The belief that breastfeeding was part of a woman's "natural" abilities also suggested that women who failed at breastfeeding were not cut out for motherhood. In the midst of bragging about her bouncing baby boy in a letter to Dr. Spock, one mother inserted her breastfeeding struggles in parentheses. "Our friends are amazed that our little boy (five months old) eats solids, is active and so wiggly and strong, and is contented with cow's milk," she wrote, quickly followed by the admission that she had "(. . . failed in breast feeding)."[28] For her, it was important that she convey to Spock that she had tried, yet failed, to breastfeed.

Aside from the institutional and psychological barriers to breastfeeding, women faced real physiological difficulties. Cracked and bleeding nipples, clogged milk ducts, infections and abscesses accompanied by screaming and hungry infants were the ingredients of a bottle-feeding household. While not inevitable these difficulties loomed large, particularly in a period when most women had little or no access to real breastfeeding support. Child experts and maternal-support organizations received countless letters from mothers throughout this period that often recounted in detail unnerving periods of illness, pain, and dwindling milk supplies as well as anxieties over infant weight and bowel movements. One mother wrote in 1953 to thank Edith Jackson for responding to her request for basic information on breastfeeding. "I wish to thank you for the copy of 'Management of Breast Feeding' which arrived in time to accompany me to the hospital. Since the hospital here as well as the doctors are uninterested in breast feeding, I felt it particularly important to seek help elsewhere."[29] Poring over the pages of the article, the mother had been able to keep herself calm during the torturous wait for her milk to come in. After a long stretch of seemingly endless days she observed: "To my surprise, my milk did come in at the end of 96 hours." Jackson's article also kept her confidence up through a "bout of mastitis" when her baby was one month old. "Fifty odd hours of chills, fever and sweating left my body dehydrated," she remembered, but again the article provided her with the information she needed to get through it. "By stimulating the breasts with a pump and excessive fluid intake I was able to re-establish my milk flow," she explained, and she had since given the article to one of her friends who had also made good use of it. "I have had such pleasure and satisfaction from the nursing experience," she wrote in closing, thanking Jackson for all that she had done.[30] Mothers who triumphed over such difficulties frequently expressed feelings of pride and happiness about the end. In a 1956 letter to Dr. Spock, one mother explained that breastfeeding produced "more contented babies and healthier and happier mothers." She went on to proudly share that she was "still nursing our youngest (our fourth)," an eleven-month-old who could drink "plenty of cow's milk from a glass but wants Momma's 'love milk' and . . . wants *no* substitute." She had delivered all four children "via Natural Childbirth" and had breastfed each of them, she wrote, "as God intended."[31]

Though Jackson, La Leche League, Niles Newton, and other breastfeeding experts undoubtedly helped many mothers through these difficulties, the vast majority of mothers would have been left to confront the uncertainties of painful and worrisome breastfeeding difficulties with little to no real medical support. In the majority of cases, mothers would have had little choice but to follow the dominant advice of most postwar physicians and turn to the bottle. One mother composed a letter to Jackson over a period of weeks in the summer of 1954, illustrating the evolution in her efforts to breastfeed and her diminishing desire to continue the struggle. She began writing two and a half weeks after the birth of her son. She had prepared for childbirth and breastfeeding with exercises and nipple-preparation techniques just as all "the books" said. After the delivery she had followed all the advice on breastfeeding she could find, including hand expression, breast massage, and hot and cold compresses to help combat engorgement. She had "been conscientious about drinking extra fluids" and was trying to get enough rest, though she admitted that she hadn't "had much sleep at night" since the baby arrived. Yet despite all her efforts, she felt undermined by her physicians who, she said, believed modern society was incompatible with breastfeeding, and she worried constantly that her son was not getting enough milk. His nursing sessions, she reported, often lasted for an hour or more and he seemed to be constantly hungry. By the time she completed the letter a week and a half later, the mother revealed that she had all but made the decision to switch to formula feeding. "At this point, formula feeding seems to be much easier for both the baby and me," she wrote. "It would relieve the tension and enable me to get my mind off nursing and on to other necessary things." Were the values of breastfeeding great enough "to make it worth a good deal of emotional strain in the family?" she wondered.[32] And yet even though she had apparently arrived at a reasonable conclusion for herself and for her family, she sought reassurance from Jackson that she was in fact doing the right thing. Despite the widespread popularity and normality of bottle-feeding in this period, women who were dissuaded from pursuing breastfeeding or forced to give it up due to logistical or medical difficulties could carry regret about the experience with them throughout their lives.

The Breastfeeding Family

The belief that an infant needed a mother and not just a caretaker is central to understanding the renewed interest in breastfeeding at midcentury. Through breastfeeding, many believed a mother could fulfill the instinctual need in both her child and herself to create a physically and emotionally intimate relationship, one that could sustain the child throughout his or her life while also allowing the mother to realize her full potential as a woman. As a 1948 issue of the popular health magazine *Hygeia* commented, "Breast feeding is a maturation point in the sequence of maternal development which is important physiologically and psychologically to mothers. . . . In this intimate situation, as in no other, a mother obtains a demonstration of her real importance to the infant."[33] As psychological arguments about the importance of emotionally appropriate infant rearing took root in these years, an understanding of breastfeeding as a necessary stage in a mother's biological and psychological development grew in tandem.

The emphasis that popular breastfeeding discourse placed on the importance of nursing for both child *and* mother was a key feature of natural motherhood ideology. While arguments about the "natural" superiority of breast milk can be found throughout medical advice literature from the early twentieth century, the benefits had been previously couched in terms of nutritional and psychological gains for the *infant*.[34] By the postwar era, experts and lay women alike were discussing breastfeeding in terms of what it would do for the *mother*, including evidence that breastfeeding hastened physical recovery after childbirth and the psychological argument that it helped "the mother feel close to her baby."[35] This perspective certainly informed La Leche League's mantra, "better mothering through breastfeeding."[36] In addition to stressing the benefits of breastfeeding for the mother, natural motherhood supported a distinct breastfeeding *family*. Both lay and scientific breastfeeding advocates from this era stressed that the benefits of breastfeeding went beyond just the mother-child pair to impact the entire family. Breastfeeding families, supporters argued, fell into a "natural" rhythm and order that appealed to those who believed that the erosion of "traditional" gender-role differences in the home were a threat to American society and the family. As the mother nourished and protected her infant, she fulfilled her own instinctual needs and desires, but the father, too, could satisfy his need to protect and support his family by warding off naysayers and shielding his wife from criticism and negativity. Breastfeeding and the science of "the natural" upon which it was built appeared to some in the postwar era as a universal salve for a host of modern "problems," a list that could encompass anything from homosexuality to divorce.[37]

Beginning with the theories of Freud and expanding over the decades of the twentieth century, ideas about the importance of both a mother and a father on the psychological development of small children helped, by midcentury, to shift expectations for what a father should be to his children and what a husband should be to his wife. By the end of this period, these trends had worked their way well into the lives and expectations of mainstream middle-class families.[38] By 1969, the *Parents' Magazine Baby Care* manual reflected on the sweeping change that had occurred when they observed: "It used to be that the husband earned the living and the wife brought up the children. . . . Now we realize how necessary it is for the father to gain the deep love and respect from his children . . . from the very beginning."[39]

An increase in the domestic demands on fathers in the postwar years and onward is reflected in the period's child-rearing advice literature.[40] Spock and others who supported bottle-feeding emphasized the benefits of the team approach to infant care that such an option allowed. With bottle-feeding, fathers could potentially share more deeply in the emotional responsibilities of infant rearing. For many breastfeeding advocates, however, this was precisely the problem: the maternal-infant relationship ought to be an exclusive one. In making this argument, those who embraced breastfeeding and natural motherhood offered an alternative to the postwar era's burgeoning ideal of the modern nuclear family.[41] In their book *The Womanly Art of Breastfeeding*, published in 1958, the League wrote, "Much confusion has taken place in many modern homes in the roles of husband and wife. . . . No wonder so many boys and girls are mixed up about their roles in later life." The answer for this "gender confusion," according to the League, was "to urge breastfeeding in order to help keep these father and mother images clear!" Their published pamphlets and internal newsletters repeatedly articulated a set of broad concerns over the importance of cultivating a specific kind of motherhood, one defined by biological difference and maintained through the cultural delineation of gender-specific spheres of authority.[42] The benefits of breastfeeding appeared to offer legitimacy to this worldview and simultaneously served as prophylactic against further erosion of maternal power in the home. "Let's see the men take [breastfeeding] over," they wrote. "As long as women continue to bear babies *and nurse them* [emphasis mine], there is not much danger that the roles of mothers and fathers will become badly confused."[43]

Concern over the blurring of distinct gender roles with regards to parenting did not necessarily translate into other rigid divisions of labor characteristic of the postwar period. In fact, the breastfeeding mother could be far more disruptive of household routines than the one who bottle-fed. In breastfeeding households fathers were often encouraged to help out by assisting with household chores.[44] One mother wrote to the League saying that while she thought their breastfeeding guide was "the best on the subject," she disagreed on their limited portrayal of the "Father's role," saying "My husband helps me and I help him. We don't say 'this is a wife's job and this is the husband's.'"[45] League leaders and other breastfeeding supporters during this era avidly discussed the relationship between infant care, maternal authority, and the role of the father in the home.[46] When mothers chose to breastfeed their infants in these years, they sought out and embraced the benefits it offered them as women.

That breastfeeding and the ideology of natural motherhood may have been, in part, a response to concerns over the widening domain of fathers in the home is further supported by the tenor of discussions about the benefits of bottle-feeding. As one woman wrote, in addition to providing freedom, flexibility, and the ability to work outside of the home for the mother, bottle-feeding allowed her husband to "share in the fun and joy" of feeding the baby.[47] Another woman recounted to Spock in a 1956 letter that she was forced to switch to bottle-feeding after three weeks of nursing resulted in severely cracked nipples and a breast infection. "From this sad incident," she reflected, her opinion shifted to being "in favor of the bottle." She recalled that her "husband took over complete care of the baby" when she became sick, and "as a result became as efficient and relaxed in baby feeding and care as any mother." She closed her letter by stating, "There is something to be said for letting father know the closeness of feeding his offspring, too."[48] Spock and many of his advice-giving peers saw no reason to push breastfeeding for every mother, particularly in light of the emotional benefits that women attributed to bottle-feeding. Bottle-feeding parents and their supporters in the medical community often highlighted that mothers could avoid the burden of bearing all of the responsibility for infant feeding. Concern for the overworked, exhausted, and irritable nursing mothers in the world, one father wrote in *Parents' Magazine* in 1950 to argue that a mother "feeling guilty about giving her baby a bottle won't do anybody in the family any good."[49] "Perhaps the natural thing *is* best," he added, "but not one of us is completely as Nature intended us to be."[50]

While the breastfeeding decision impacted negotiations over family organization, gender roles, and parental authority, it also affected power dynamics when it came to sexual relationships. The contested nature of the female breast as both a symbol of sexual desire and one of maternal nurture became uniquely problematic within the context of the era's scientific and popular ideas about sexuality and family. The need to reconcile an increasingly sexualized breast with a maternal breast surfaced as a struggle in the discussions and practices surrounding breastfeeding. As one advice author wrote, "It is generally recognized that some young husbands object to their wives' breast-feeding; the breast is a sex symbol that they regard as their own exclusive property."[51] Although certainly the duality of the female breast was not a new phenomenon in the postwar era, historians have argued that the sexual obligation of women in this period took on a different and more intense character than that of previous eras.[52]

Proponents of natural motherhood attempted to diffuse the sexual tension surrounding breastfeeding by highlighting that maternal activities like childbirth and breastfeeding were necessary parts of a healthy and well-rounded expression of female sexuality. Most notably, psychologist Niles Newton and her husband, obstetrician Michael Newton, demonstrated in 1948 that there were physiological links between intercourse and breastfeeding in the body's production of the hormone oxytocin. When science journalists Ruth and Edward Brecher reported on the Newtons' work, however, they received a flurry of outraged letters in response to their description of breastfeeding as "sensuously satisfying."[53] As Edward Brecher later recalled, "Recognition of the sexual nature of breast-feeding . . . is deeply troubling and anxiety-arousing for many women . . . 'How dare you say that there is anything sexual about so warm and loving an act as nursing a baby!' one of our readers wrote us indignantly."[54] The idea of a female sexuality expansive enough to encompass components of a woman's life beyond the marriage bed threatened the fragile postwar social order. Advice books, journals, and magazines attempted to understand the nature of the relationship between breastfeeding and a woman's heterosexual obligations, providing evidence and stories of nursing mothers' sex drives and body shapes returning more quickly or more slowly than bottle-feeding mothers.[55] The continued exploration and discussion of such issues suggests that in negotiating these sometimes-uncomfortable topics, mothers confronted the question of what it meant to inhabit a sexual and maternal body.

Competing Desires

The proliferation of mass media such as the pinups of World War II and of *Playboy* magazine, launched in 1953, helped to crystallize an increasingly hegemonic male-centric female ideal, one that emphasized the male-female sexual relationship as the foundational and primary relationship of the postwar social order. In this sexually heightened atmosphere, the male gaze came to focus intensely on the female breast, eroticizing this part of the female body more so than in previous periods.[56] While the interwar years had witnessed innovations in women's fashion technology that emphasized the female bust, either its flatness or its curve, by the postwar period, variation in breast shape and size was no longer desirable. The quest for a standardized, voluptuous, and sensuous breast consumed many men as well as women, spawning a whole new era of breast anxiety.[57] The problem was serious enough to be the subject of a meeting of gynecologists in 1955 during which a physician from the University of Oregon claimed that women were becoming "neurotic" over the size and shape of their breasts.[58] As historians of women's fashion Jane Farrell-Beck and Colleen Gau have noted, by the end of World War II the desire for ample bosoms combined with technological advancements in undergarment construction to give birth to the mass production of the youthful conical breast, epitomized in so many ways by the infamous costume and silhouette of the *Playboy* "bunny," launched in 1953. In one of *Playboy*'s earliest issues, an advertisement insert for a subscription to the magazine featured a woman dressed in bunny ears and a form-fitting bodysuit, which covered her torso from the neck down, except for two circular cutouts in the fabric exposing and emphasizing each individual breast.[59]

Scholars have suggested that the shift of the heterosexual male gaze to the female breast during the postwar era may have played a significant role in the decline of breastfeeding during these years. Sociologist Linda Blum has observed that in these years "when breasts were singled out and increasingly sexualized," breastfeeding was "doubly dangerous." Nursing, she has argued, "threatened to expose the breasts to the heterosexual gaze, but also to compromise the object of that gaze"; thus breastfeeding both revealed the breast and blatantly claimed it for maternal use.[60] The centrality of the breast as a symbol of compulsory heterosexuality during the postwar years led to a complex web of social constraints and internalized discomfort over breastfeeding.[61] As historian Marilyn Yalom has argued, the resonance of the female breast as a symbol of comfort to soldiers during World War II lasted into the postwar era as the men returning home "needed to be reassured that the nightmare of war was over and that the breasts they had dreamed of were now available to them." The heightened cultural awareness of breasts as both maternal and erotic symbols also sent a clear message to American women in the postwar period that their role was "to provide the breast, not the bread."[62]

At the same time that the female body, and particularly the female breast, came under increased public attention, the pronatalism of the period offered the potential to challenge this particular reconstruction of the breast through the ideology of natural motherhood, of which breastfeeding and maternalism were of the utmost importance. La Leche League, quite aware of the sexual tensions that surrounded the question of breastfeeding, wrote to reassure mothers-to-be that the marital breast and the nurturing breast could coexist: "Most [husbands] . . . are enthusiastic about the idea from the start; they know without much thinking about it that it is not only the best way but the *womanly* way to feed a baby; and womanliness is the trait they most value in a wife."[63] While popular culture and media stressed the sexual importance of the female bosom in male-female relationships, breastfeeding advocates offered an alternative model in which the maternal, nursing woman sat at the locus of male desire. According to the League, true men desired "womanliness," including pregnancy and motherhood, not the perky breasts displayed by *Playboy* centerfolds. This construction of female sexuality and marriage during this period was clearly an alternative view, one which many women had a difficult time adopting given the strength of the America's cultural obsession with breasts. Even Spock wrote as early as 1946 about the connection between breastfeeding and sexuality: "Some mothers shy away from breast feeding for fear that it will ruin their figures. . . . As for the effect of nursing on the shape of the breasts, I am sure that, in many cases, it causes no permanent change. . . . After all, the breasts will sag from becoming too fat even without pregnancy or nursing."[64] Spock was not against breastfeeding, but for very particular reasons saw benefits to bottle-feeding that many staunch pro-breastfeeding advocates did not.[65] In a popular scientific magazine article, Betsy McKinney argued that while in most cases husbands did not actively sabotage breastfeeding, "research [has] indicated that husbands are sometimes the reason for their wives' inability to nurse." McKinney elaborated: "With the birth of their child, her breasts become practical, working equipment, no longer romantically reserved for him alone, but plainly used for the satisfaction of another." This "new" purpose for his wife's breasts could lead to "confusion" that could be "impossible for him to resolve." Although she suggested that there were "many men who view their wives' nursing participation with whole-hearted happiness," she also acknowledged that many found "this function so shocking" that they could not "even observe the nursing relationship without feeling ill at ease." In short, McKinney reported that it would not be abnormal for a husband to find it "difficult . . . to share his wife's person so freely," even with his infant.[66]

The tension over the sexual and nurturing potential of the female breast was also highlighted by Alfred Kinsey's 1948 study on male sexuality. The report found that for nearly all heterosexual male respondents the female breasts were primary objects of sexual desire. He further revealed that over 96 percent of heterosexual men surveyed reported that breasts played a vital role in their sex lives.[67] In his 1953 study on female sexuality, Kinsey found that a majority of women refused oral-breast contact until after marriage, or at least until after they were "more experienced." Furthermore, the study indicated that women felt that "the acceptance of any oral technique involves something that is peculiarly sexual, or that it represents an extreme break with the moral traditions of our culture."[68] In a period characterized by an expanding culture of premarital sexual practices, Kinsey's work suggests that access to the female breast occupied a particularly meaningful spot in the hierarchy of dating activities.[69]

There is suggestive evidence elsewhere, as well, that women thought about the tensions between the sexual meaning of their breasts and their interests in breastfeeding. In a series of interviews with new mothers conducted by the New York City Health Department in 1952, researchers sought to investigate women's attitudes toward breastfeeding. Twenty-three percent of those interviewed reported that breastfeeding was embarrassing and "offended their sense of propriety." One woman interviewed went so far as to comment that she felt it to be "animal-like."[70] After witnessing a television program on which Dr. Spock had discussed the many aspects of breastfeeding, including, apparently, the subject of marital relations, one mother defiantly wrote: "With regard to the figure, my insignificant bosom has become almost concave—whether a natural result of the passage of time or because of breast feeding I don't know—but my milk is abundant and my babies fat and firm without solids or supplementary feedings. It seemed to be all the more necessary to be functional since I was not decorative in that department."[71]

Concern over the aesthetic effects of breastfeeding on the breasts frequently surfaced in discussions concerning the importance of nursing bras as necessary technologies for maintaining the breasts in their pre-nursing state. One mother wrote to Spock in detail about her bra issues during pregnancy. "I wear a B to a C cup now in a normal bra," she wrote, "but if I buy a nursing bra, I have to get a D cup . . . and it is still too small now." With a sense of exasperation, she added, "Is there nothing that can be done for this?"[72]

Women's magazines, newsletters, and correspondence from the era suggest that, for many women, feeling sexually desired by their husbands was an important part in their experiences with breastfeeding. The fact that a notable percentage of women succeeded in breastfeeding, however, indicates that despite the ideals articulated in popular and scientific discourse, women and their families found ways to combine the seemingly disparate roles of the maternal and sexual breast. In the process of going about their daily lives, breastfeeding women constructed an experience of motherhood that worked for their families. Having breastfed three children for nine months each and nursing a fourth, one mother wrote to Spock in 1957 to share her opinions. Confounded by the lack of support for breastfeeding in the medical community at large and by society's attitude toward women's breasts, she wrote, "It's fine for women's bosoms to be exposed on T.V. and radio and magazines, but the idea of a mother nursing a baby in front of anyone but her husband or mother is shocking." For her, breastfeeding was not only enjoyable; it was a source of pride. She boasted, "My husband thinks it's the most beautiful sight in the world." When it came to the rest of her family, she had included them in the process as well, writing "the older children (four and six) wanted to know what my milk tasted like [so] I put some in a glass for them. Their opinion—coconut milk."[73]

Aside from concern over protecting the nursing breast so that the marital breast might someday return, the process of preparing the breasts for breastfeeding began long before the infant's arrival. All advice manuals provided direction on how women could prepare their breasts for the task of breastfeeding—a process that, if done properly, might begin months before the baby's birth. The rituals themselves that surrounded the preparation of the breasts, both before and after the birth, were involved, time-consuming, and helped to mark the change in function that the breasts would be undergoing. Women were warned that if they wanted to breastfeed, they should start toughening their nipples and sizing them up for appropriate shape, size, and position as soon as they learned of their pregnancy. An article in a 1949 issue of *Parents' Magazine* explained these "simple" procedures as follows:

> Care of the expectant mother's breasts consists fundamentally of a simple technique of daily washing of the nipples with plain soap and water, employing a rough, clean washcloth. Before beginning such a procedure, the hands should be washed thoroughly to prevent any infection. The dark portion of the breast surrounding the nipple should be held with one hand so that the nipple protrudes. This leaves the other hand free to apply soap and water by means of a rotary motion of the washcloth. The right breast is held by the right hand and the left hand does the cleansing. In like manner, the left breast is held by the left hand and the right hand does the cleansing. After rinsing

> off the soap with plain water, each nipple should be dried with a rough towel and some type of lubrication applied with the fingers.[74]

Experts wrote detailed instructions like this to help women prepare and "toughen" their nipples so that the strain of a sucking baby would not be as traumatic, but there was also a degree to which these processes helped distinguish the maternal breast from its non-maternal functions. Familiarizing a woman with a routine like this, for example, also helped her get used to the idea and the sensation of having an infant suckle at her breast. It no doubt also helped mothers-to-be to see and think about their breasts in a new light. As the Children's Bureau itself suggested, "Taking good care of your breasts will help in making breast feeding successful. From the start wear a brassiere that gives them good support. Bathe them at least once a day, and clean off your nipples after each nursing."[75] Marking the breasts for nursing long before, and after, the birth of the infant was one solution to the breast's heightened sexualized function in the postwar years.

Despite the lengths to which advice authors went to explain the intricate necessities of distinguishing the nursing breast from its other roles, they occasionally succeeded in doing just the opposite. As Niles Newton wrote in her monograph, *Maternal Emotions*: "When mother and baby are undressed they press firmly together, skin against skin. This kind of firm, continuous pressure is described by Kinsey, et al. as leading to sexual excitement."[76] While many women who embraced natural motherhood and breastfeeding may have found these connections liberating, it also undoubtedly made some women and their sexual partners feel uncomfortable with the idea of breastfeeding. A growing popular understanding of Freudian theories linking infant experiences to sexual development later in life may have also caused mothers concerns when it came to breastfeeding. Many letters addressed to Dr. Spock during these decades expressed parental anxieties over children as young as fourteen months of age displaying what they believed was a type of sexual awareness. One mother wrote to Spock in great distress over her nineteen-month-old boy whom she described as making "up and down motions" on top of his blanket and stuffed animals. "I didn't know there were sex feelings that early," she wrote, "and I just don't know what to do about it."[77] Her son's behavior apparently caused her such anxiety that she dreaded having company over and was wary of leaving him to play with his older sister lest he corrupt her. With some breastfeeding advocates suggesting breastfeeding should last well beyond the first year of life, observations that young children might possess "sex feelings" alarmed some. Another mother wrote with a similar concern about her seventeen-month-old daughter who "about six months ago . . . discovered that she felt some sensation from her organs." "She crouches down on her hands and knees and moves back and forth," she wrote, adding, "You can see that she is getting some sort of a satisfactory feeling from this."[78]

Mothers worried not just about the sexuality of their children in infancy, but more so about their adult proclivities. Mothers in the postwar years could be keenly aware of a responsibility to raise heterosexual sons, and increasingly heterosexual daughters. Historian Donna Penn has argued that the 1950s was a period of heightened concern over a perceived increase in lesbianism specifically. Popular conceptions of sexuality in this period tied sexual expression to gender roles, which meant that the expression of acceptable forms of femininity and masculinity was central to the development of sexually "normal" children. The widespread belief that homosexuality was on the rise in the postwar years led experts and lay people alike to blame changes in the culture of marriage and the family for producing "confused" children. Also by the 1950s, she writes, science had concluded that homosexuality was a psychological state, a shift from earlier theories, which had posited it as a glandular and hormonal issue.[79] These expectations combined to raise flags of concern for many mothers who observed their children's sexual development with great awareness and anxiety. One mother wrote to Spock, in veiled yet not-inscrutable language, of her concern over how to raise normal, "productive children." She wrote, "How does a mother (or father), who knows that he or she is of a naturally dominating temperament, avoid dominating their children? I know all too well the evil effects that can result, and as the mother of two sons, I am very much concerned about this. I want above all things for my sons to grow up free to be themselves within a framework of rules imposed by God and society."[80]

While many scientific experts and maternal advocates may have agreed "breast was best," there was little agreement to be found in how to address questions such as: When should a child be weaned? What was normal behavior for an infant and what was abnormal when it came to behaviors that appeared sexual? If a mother breastfed her infant son for too long, was she "over dominating" him and inadvertently turning him into a homosexual? Concerns like these stressed the breast as a site upon which cultural anxieties about sexuality were made manifest.

No matter how a mother chose to mediate the maternal and sexual aspects of her breasts, active engagement in this act of self-determination through breastfeeding could bequeath a sense of self, purpose, confidence, and awareness that many found exhilarating. After receiving guidance from her local La Leche League group, one mother wrote of the joys of successful nursing, "After three bottle-fed babies I am finally a successful nursing mother. The striking difference is the tremendous feeling of satisfaction that this baby has given me. Up till now I have been always completely baffled by the phrase 'enjoy your baby.' Not so with my breastfed baby. She, at three months, and I are already great friends. I feel as though I have finally arrived at motherhood."[81] Others shared similar experiences, expressing feelings of "completeness, of fulfillment" that they had not experienced with their bottle-fed babies.[82] Not all women who breastfed found refuge in a La Leche League group, however. Given the significant numbers of women who attempted breastfeeding throughout this era, it is likely that there were success stories like these happening across the country through local childbirth education classes and other breastfeeding support groups like the Boston area's Nursing Mothers Council. This widespread active engagement with the "natural" maternal aspects of their sexuality helped women like these discover a sense of self that subverted the era's ideologies of containment and scientific motherhood.

By the early 1970s, the resurgence in breastfeeding was well under way, and American women were becoming far more comfortable with breasts that both nourished and provided sensual fulfillment. One mother in writing to the League's Medical Advisory Board in 1970 discussed in a matter-of-fact way how she and her husband dealt with their infant daughter's abrupt self-weaning at the age of "11 ¾ months." "Naturally my breasts became full," she wrote. "I had extreme difficulty in expressing milk from my right breast, and as a last resort, at 2a.m., my husband nursed it."[83] The struggles of second-wave feminism also alleviated some of the tension over breastfeeding. With the new generation's focus on equality in the public sphere, insecurities over maternal authority in the household were replaced by disputes over gendered divisions of labor, legal parity, and placing checks on the authority of medical expertise. By the 1980s, widespread availability of breast pump technology helped to loosen and expand the breastfeeding pair to include fathers, grandparents, and even paid caregivers. In the closing decades of the century, the breastfeeding question acquired a new set of meanings and implications, but as I have argued, our modern interest in breastfeeding first reemerged as part of a construction of natural motherhood that challenged postwar gender roles within the family.

Roots of Modern Breastfeeding

Postwar mothers navigated a changing landscape when it came to family life, marriage, and breastfeeding. When Eleanor Lake wrote in *Reader's Digest* in 1950, she breezily drew a connection between breastfeeding and that cornerstone of Victorian women's wear, the bustle. That she did so suggests more than just a rhetorical flair. The movement back to the breast that Lake helped chronicle symbolized an important connection to an earlier era.[84] Lake, however, did not go so far as to suggest that women bring back the bustle along with the breast. Rather, for Lake and for countless other women in the postwar years, breastfeeding offered the possibility of a bridge between two eras—a practice that promised to unite the moral legitimacy of the nineteenth-century mother with the twentieth century's modern and scientific worldview. The ideology of natural motherhood appealed to some women in this period who sought to imbue their maternal identities with a degree of purpose, moral authority, and sexual autonomy that modern motherhood and postwar housewifery threatened to extinguish. Mothers who pursued breastfeeding in these decades resisted the increasingly reductionist and largely meaningless biological definition of "Mom" in these years, which positioned mothers as simply a circumstantial member of a group of "female individuals—some admirable, some not—who had been through a certain biological experience."[85] In doing so, these women helped popularize a new scientific and popular ideology of motherhood and breastfeeding that stressed nature as a source of legitimacy and moral authority.

The ideology of natural motherhood also offered a potential challenge to the growing hegemony of the middle-class housewife, a role that posited women primarily as consumers of domestic goods and sexually accessible partners to their husbands. As one Ohio mother and La Leche League member put it in 1959, "I used to believe what Dr. Crane [advice author] and others write in their columns about, 'If your husband

and new born baby call you at the same time, go to your husband.' But I do not believe this any more. If you are married to a big baby and are afraid of losing him, then go to him. But if you have a grown man as father of your baby, your first duty is to the baby."[86] Breastfeeding mothers upset the postwar order when they put their infants' needs above those of their husbands and when they privileged the maternal breast over the sexualized breast, if only temporarily.

In embracing natural motherhood and breastfeeding, however, women also continued to operate within a system of historical limits and political constraints. Choosing between an all-consuming maternal identity and an all-consuming role as a housewife clearly limited women's authority to the home and fell under the era's dominant ideology of containment within marriage. Women's personal struggles for control over their infant feeding choice, their bodies, and their families, however, also provided some breastfeeding mothers with the opportunity to experience a sense of empowerment, pride, and even the awakening of a sense of gender consciousness that helped lay the groundwork for the surge in breastfeeding and maternal health activism of the late 1960s and 1970s. Perhaps most importantly, however, maternal health and advocacy organizations helped build networks of female-centric expertise that mothers utilized to obtain the kinds of maternal experiences they wanted. Mothers across the country helped disseminate knowledge about breastfeeding to their friends and families, and they also exchanged information about physicians and hospitals in their communities that were supportive of breastfeeding. Although women who breastfed successfully for more than a few weeks represented only a small percentage of American mothers overall, they were instrumental in bringing about a sea change in infant feeding practices by the 1970s.

CHAPTER FOUR

Maternal Expectations

New Mothers, Nurses, and Breastfeeding

Working in the 1950s, Muriel McClure was a nurse, and a realist. "Breast feeding can be a happy and satisfying experience for the new mother and her baby," she wrote, adding, "*if* she gets the support, encouragement, and instruction she needs in her first attempts at nursing."[1] Writing in 1957, McClure was also a member of an elite group in a number of ways: she had earned a master's degree in nursing, was a published author, and had happily breastfed her own two daughters. Her experiences as a maternity nurse in New York Hospital's Rooming-in Division had given her additional expertise and legitimacy as an expert on breastfeeding. By the 1950s, a handful of nurses like McClure pushed to change how their profession handled breastfeeding. Ten years later, during a workshop on breastfeeding held for nurses and students in 1967, instructors Loraine Dyal and Julia Kahrl set out to tackle some of the obstacles that continued to prevent nurses from being effective breastfeeding supporters. The most pressing item on their agenda was to help the nurses work through their inhibitions and discomfort with their breastfeeding patients. Dyal and Kahrl encouraged the nurses to "evaluate their own attitudes toward breast-feeding" and asked them to consider their own personal history with the practice—had they been breastfed? Had breastfeeding been a part of their family experience as a child?[2] By the end of the 1960s, more and more nurses noted that they sat in a position of unique authority and power when it came to breastfeeding. Dyal and Kahrl hoped to encourage the next generation of nurses to embrace their own maternal experiences (whether as mothers themselves, or as daughters, aunts, friends, etc.) as integral components in their practice as caregivers to breastfeeding mothers.

Throughout the postwar years, many physicians and nurses alike accepted that breastfeeding was best, in theory. Furthermore, by the 1950s studies from ethology, psychology, and maternity rooming-in wards had helped demonstrate that the best way to get a mother to successfully breastfeed was through calm and consistent encouragement and support from the period immediately after birth throughout the first weeks and months of new motherhood.[3] Providing this kind of support in the hospital overwhelmingly fell on the shoulders of the nurses, presenting both a laborious burden as well as an opportunity for greater authority within the realm of postpartum and pediatric hospital care. In practice, however, mothers' demands for breastfeeding support more often than not provoked discomfort, embarrassment, insecurity, impatience, and misunderstanding from their nurses throughout the postwar decades.

From the postwar years through the early 1970s, the United States experienced extremely low breastfeeding rates and nurses, much like doctors, received little to no formal training on how to assist breastfeeding mothers effectively.[4] Made invisible by their nature as caring laborers and increasingly cast as simply another component of the technology of the hospital in which they worked, nurses have often been overlooked by scholars who have explored the history of breastfeeding.[5] Analyzing the role of nurses and the labor of caring within the structural hierarchy of the hospital continues to pose challenges to historical investigations of nursing and health care. Yet, through a postpartum mother's eyes the nurse was an all-important figure: the gatekeeper to her infant and to her dreams of what kind of mother she would be, a breastfeeding mother or a bottle-feeding mother? For the nurse, this influence might have often been unwelcome, unnoticed, or alternatively, some nurses might have actively wielded their influence to favor breastfeeding over bottle-feeding, or vice versa.

Women's writings and nursing publications from these years provide clear evidence that the structure and policies of the hospital shaped infant feeding interactions between nurses and mothers to a significant extent. As individuals, however, nurses struggled to meet the competing demands of their patients as well as those of their employers while mothers sought to exert agency over the care and feeding of their infants.[6] Nurses not only brought their scientific and professional training to bear on their interactions with mothers; they also harbored their own personal beliefs about motherhood and infant feeding, often informed by their position in society as much as by their individual experiences. This was particularly true in the case of postpartum care in a period when fewer and fewer physicians devoted much time or attention to the process of breastfeeding. The hospital's overall indifference to breastfeeding throughout this period in particular provided nurses with a space in which they could exert greater professional authority by taking charge of breastfeeding mothers. Alternatively, it also created room for mothers to negotiate, plead, or even manipulate nurses and doctors into giving them what they wanted, in this case, to breastfeed. As some mothers struggled to exert individual agency within the limiting confines of the hospital environment, the maternal breast once again served as a battleground on which much larger disagreements over ideology, personality, and professional identity could be waged.

The Postwar Hospital

By the end of World War II, childbirth in the United States had become an occasion for the routine hospitalization of healthy and robust young women. In the context of the postwar hospital, women's birthing and early motherhood experiences were defined and structured by the latest in scientific efficiency, aseptic standards, anesthetics, and analgesics. By 1950, 88 percent of all births in the United States took place in the hospital and in little more than a decade, having a baby outside of the "sterile, managed delivery room of a hospital was considered dangerous and foolish and was practically nonexistent."[7]

The midcentury turn toward hospital birth in record numbers was no mere coincidence. In addition to the "baby boom" of the postwar years, the decades following World War II were uniquely oriented toward the growth of scientific and medical institutions. Fueled by the boom in research and technology wrought by America's war programs, the 1946 Hill-Burton Act (also known as the Hospital Survey and Construction Act) ensured the exponential growth of the nation's community hospitals, laying the structural groundwork for the rise of the hospital as the center of the American health care system.[8] As Rosemary Stevens has observed, "The belief that the hospital could be both a technological and a community institution, with little conflict between the two goals, reached its zenith in the fifteen years after World War Two."[9]

With more mothers entering the hospital than ever before, opportunities for exposing women to the scientific methods of infant care multiplied. Such methods almost universally relied upon standardized formula feeding of the newborns that inhabited the sterile rows of beds. Stacked behind the iconic viewing glass through which anxious fathers and other family members could see—but not touch—their "new addition," these infants routinely received carefully timed and measured formula feedings from the nurses each day.[10] From the perspective of a nurse in this period, the breastfeeding mother was an anomaly at best and at worst could provoke the ire of the nursing and medical staffs for upsetting the efficient flow and order around which the hospital had been built.

Given the increase in hospitalized birth and the decrease in the practice of breastfeeding in general, the recorded interactions of nurses with breastfeeding mothers offer a unique historical opportunity to understand, with a rare degree of intimacy, what women's "choice" to breastfeed meant in those first hours and days after birth. Exploring the nature of these interactions requires that we carefully examine the ways in which nurses and mothers experienced the brief moments when a new mother first put her infant to her breast.[11] Accounts by mothers about their nurses remained significantly consistent throughout this diverse time period.[12] Many would-be breastfeeding mothers, upon familiarizing themselves with the hospital setup, quickly learned to give up any expectations of finding allies on the nursing staff. Those who were most prepared for the uphill battle they would face in the hospital went so far in some cases as to obtain written instruction from their physicians weeks in advance of their due date, prescribing how staff should care for the breastfed infant.[13] However, even in these optimal scenarios of educated, prepared mothers and supportive doctors, institutional inertia, of which the nurses were a part, often prevented women from achieving their ideal breastfeeding experience. As one 1950 article in *Ladies' Home Journal* pointed out, "fifty percent of mothers 'don't want to,'" and "half of those who do try, fail."[14]

Nurses, in following hospital procedures, gave infants sugar water or formula regardless of whether or not the mothers intended to breastfeed. Additionally, nurses followed hospital policies of feeding infants on three- or four-hour schedules, which meant infants who were breastfed often ended up screaming with hunger for hours in the nursery only to arrive to their mothers at "feeding time" fully exhausted and asleep.

Despite the growing faith in the "demand" feeding schedule in theory, hospitals found it logistically difficult to do anything but adhere to a strict set of guidelines and rules when it came to feeding infants.[15] Nurses, meanwhile, complained that breastfeeding mothers cost them extra time and patience and that even in cases when hospital policies allowed for breastfed babies to be fed breast milk exclusively, nurses—in an attempt to quiet their unsettled charges—would often give them formula anyway.[16]

The degree to which the maternity and nursery nurses' actions were structured by the policies of the hospital, particularly in the case of infant feeding, is notable. Women who embraced natural motherhood vacillated between understanding the nurse's limited ability to change the system and feeling disgust, anger, and frustration with the seemingly minor effort it could take on the part of the nurse to simply take the mother's feelings and wishes seriously. Given these constraints, the relationship between the nurse and the mother was fraught with tension and could go either wonderfully or disastrously depending upon the specific circumstances and the particular beliefs of the people involved. Within this atmosphere of heightened emotions, minor shifts in a nurse's attitude, action, and tone could have significant impact on a mother's breastfeeding experiences.

The Breastfeeding Encounter

In 1947, a journalist (and mother) published an article in the *American Journal of Nursing* (*AJN*). The author described what would become a commonplace description of a nurse's interaction with a breastfeeding mother. After months of encountering medical, government, and popular advice explaining all of the benefits to breastfeeding, Letitia Sage expected nothing less than nursing bliss once her infant was born. "My first-born had not yet chalked up a full existence when she was brought in and handed to me," she recounted. "'Try nursing her for five minutes and then give her this,' said the nurse, putting a bottle filled with two ounces of formula within reach." The nurse left, and Sage set to work, only to find that her daughter was "distinctly apathetic" and that she, herself, was "terrified" throughout the experience. "After a brief and unsatisfactory struggle for all concerned," she wrote, "I reached for the bottle." Her hungry daughter drained it. This process continued for days while Sage was in the hospital; each time she tried to breastfeed, her daughter's fits of howling would bring the nurses running with reassuring smiles and bottles in hand. As one of her nurses explained after a particularly tense session, "'Cheer up—she won't starve even if you can't nurse her!'"[17]

Sage's story highlighted her attempts to breastfeed alongside the actions and efforts of her nurses. Sage did not appear to think of the nurses as particularly unfriendly, or even purposely unhelpful. Rather, in her account, Sage and her nurse seemed to experience the situation of her stressful attempts to breastfeed from two very different perspectives. Sage, who was self-educated on the benefits of breastfeeding and devoted to the idea of giving her child what she believed to be the "very best," experienced nothing short of a tragic maternal failure on her part. The nurse, however, seemed remarkably unconcerned. The nurse's primary focus, rather, seemed to be that Sage was not in danger of physical harm and that the baby received basic nourishment. Formula, the nurse asserted, was entirely capable of fulfilling the infant's needs.

Despite her early difficulties, Sage reported that she did go on to eventually breastfeed her child successfully, but only *after* she had left the hospital and suffered weeks of frustration and self-doubt. She never forgot how important the nurses had been in her experience and in her essay she left her readers with some final thoughts: "Here, it would seem, is a major challenge for the nursing profession, particularly for those who are in the hospital maternity divisions or who frequently accept the care of new babies in private practice. It takes patient education to establish breastfeeding in many cases, and it is the nurse who can best give such education to new mothers."[18] Sage's tale of caution and advice presents a picture of divergent expectations about the postpartum and infant feeding experience in the hospital. No matter how much they had prepared for breastfeeding, mothers consistently looked to their nurses for help when they were in the hospital. While they sought the approval and permission to breastfeed from their physicians, mothers looked time and time again to their nurses to help them turn their desire to breastfeed into a reality. Nurses in general, however, were far from well-trained experts on breastfeeding and in fact knew just as little about its actual practice as most other medical professionals at the time. Despite scientific advances in understanding breastfeeding physiology, physicians and nurses all suffered from a lack of working knowledge on the subject. This gap between the knowledge and practice of breastfeeding plagued the relationship between breastfeeding and the nursing profession well into the final years of the twentieth century, a situation that arguably persists to this day.[19]

Assisting mothers with breastfeeding could often be a hands-on job. Nurses who tended to breastfeeding mothers often had to physically help the infant "latch on" to the mother's breast. Such efforts involved manually manipulating the breast, perhaps for an uncomfortably extended and consistent period of time. Furthermore, breastfeeding mothers required education on techniques like expressing milk from their breasts, since this could be crucial in preventing and alleviating breast engorgement. Manual expression was also helpful for enticing the infant to latch on and to stimulate milk flow and production.[20]

Nurses who oversaw breastfeeding mothers, therefore, were responsible for manually expressing milk from engorged breasts, teaching mothers techniques for doing it themselves, and perhaps even showing them how to express milk with a handheld breast pump.[21] These were banal practices that took place without a draping sheet and without some of the more formal trappings of medicine to mask the intimacy of these tasks (as, for example, a gynecologist or obstetrician might have with stirrups, sheets, and medical instruments). Not surprisingly perhaps, according to the accounts by mothers and by published discussion among nurses themselves, these practices were not typically well liked by nurses. In fact, the vast majority seemed to avoid breastfeeding mothers whenever possible and generally seemed to detest these interactions, even as many mothers, eager as they were to breastfeed, accepted them as necessary and worthwhile.[22] That there was a persistent gap between what a breastfeeding mother needed and wanted and what a nurse would actually do is further evidence that nurses and breastfeeding mothers commonly held vastly different expectations and ideas about breastfeeding and motherhood.

To aspiring breastfeeding moms and natural motherhood advocates, nurses were so commonly encountered as hindrances to breastfeeding that La Leche League devoted several pages of reassurance for mothers on the subject in *The Womanly Art of Breastfeeding*. They assured would-be breastfeeding mothers that even if their efforts to nurse in the hospital were thwarted, they would still have a chance to reestablish their milk once they returned home. "Not all hospitals are ideally set up to be of the greatest possible help," they warned. "In such hospitals you can expect a few raised eyebrows from some hospital personnel when they find out that you've 'taken the notion' to nurse your baby." The League always stressed the importance of being kind and patient with nurses; they suggested that it might be helpful to just "casually mention to the nurse that you're glad your baby isn't getting any formula—since you'll be completely breastfeeding him." It was commonly accepted that most nurses either did not know or did not care about breastfeeding. As the League put it, "the nursing staff may not know . . . about allergies, and through busy forgetfulness, or simply out of the goodness of their hearts, they may think they are doing your baby a kindness by giving him a bottle or two during those first days before your milk comes in."[23]

La Leche advocated that women should try to get their doctors to agree to let them have their babies brought to them "at least every three hours" and "certainly for at least one night feeding."[24] Adding these hiccups to a nurse's already busy routine would have no doubt created tension between the mother and the nurse, no matter how well meaning both parties may have been. Mothers ultimately knew, however, that in order to get what they wanted they often needed to turn to their physicians. In doing so, they subverted any kind of alliance they might build with their nurses. Furthermore, choosing to "go above their heads" could engender a backlash from the nurses that could make the mother's hospital stay less than enjoyable. And yet, despite this risk, well-prepared mothers who wished to breastfeed often saw very little alternative: the nurses did not have the authority to buck the hospital routine; only the physician could implement an infant feeding order, and even then, there was no guarantee that the nurses would follow through.

Barriers to Breastfeeding Support: The Nursing Routine

Nurses' subjection to hospital routine often made it very difficult for even the most dedicated breastfeeding allies to effect any measurable change. One particularly powerful institutional impediment to breastfeeding was the "dry up" pill, a synthetic estrogen named diethylstilbestrol, which many mothers received to inhibit their lactation and/or manage engorgement in the hours and days after birth. Mothers often received this medication without their complete knowledge, particularly through the early 1970s. While it was the obstetrician who left the orders for prescription drugs to be given, it was the nurses who physically handed them out.[25] As one physician put it in 1970, "Suppression of lactation and use of bottle-feeding seem now to be preferred by the majority of women in the United States," so little effort was made in most cases to accommodate mothers who wanted to breastfeed.[26] While mothers might feel friendly and even supported by their nurses during their interactions, in following such orders nurses often acted in direct opposition to their patient's desires. When asked about any feeding problems she was having in a follow-up questionnaire by the Boston Association of Childbirth Education, one mother summed up the situation when she wrote, "One nurse gave me two drying up pills and told me they were vitamins—however we settled that misunderstanding quite nicely."[27] Not all mothers would have been so well informed or outspoken. Despite widespread anecdotal evidence that nurses were, at best, withholding information about the dissemination of dry-up pills, League leaders offered reassurance and did not publicly portray nurses as unfeeling. However, they were adamant when they cautioned mothers to be wary of anything and everything the nurses did. "If you should be given shots or pills to 'dry you up,'" they cautioned, "don't worry about it, though you might try to avoid them. These are usually a routine carried over from the mother who doesn't breastfeed."[28]

The League and breastfeeding advocates in general encountered the dry-up medication phenomenon quite frequently. "If you continue to nurse your baby," they wrote, "it will not hinder your milk flow . . . with or without it, you can nurse your baby."[29] But despite reassurances like these from League publications, many mothers attributed their breastfeeding failures to the actions of their nurses. Furthermore, fighting the hospital routine was something that fatigued even the most well-educated and well-connected mothers in the hours and days after delivering a child. For the majority of women who came into hospitals with little preparation or support, bucking hospital procedures and the authority of the nurses seemed unwise and undesirable.

Women shared their woeful stories of experiences with nurses who were unhelpful, neglectful, or actively harmful during their attempts to breastfeed. These stories persist from the early 1950s (and earlier) through the mid-1970s. Despite the positive changes that hospitals were undergoing in maternity ward policies in the same era, as Judith Walzer Leavitt has shown in her work *Make Room for Daddy*, the intransigence of infant feeding policies is notable.[30] One woman, interviewed by the *Ladies' Home Journal* in 1950, discussed her struggle with trying to breastfeed while in the hospital. "The first time my baby was brought in at six A.M., the nurse poured out a rapid fire of instructions and left the room before I could ask her any questions," she reported.[31] But the young mother, clearly unprepared for the hospital onslaught against her desire to breastfeed, attempted to get by on her own. Unfortunately, after five days of going along with no further help or advice from her nurses, her doctor arrived to tell her that things were not going well and that she would have to formula feed. She described feeling "crushed" and fell into absolute despair for the remainder of her hospital stay. For this woman, the nurse was more noticeable for her absence than for anything else, but this alone was enough in this woman's eyes to contribute to her failure.

Another woman, wrote in to *Parents' Magazine* in 1952, to explain how she had felt out of place in the hospital because her desire to breastfeed seemed to cause everyone around her such inconvenience. She wrote:

> When our first baby was born, I was convinced of the importance of breast feeding. Unfortunately, the hospital I went to wasn't equally sold on the idea or, if it was, it hadn't seeped down to the semiprivate maternity floor. The nurses' attitude seemed to be, "We could manage these babies so much better if only their mothers didn't insist on horning in." I was one of the few mothers who persevered and whose baby started life on his natural food.[32]

Still another woman, who gave birth to her first child in 1960, recounted to me the following about her experience with her nurses:

> I told them that I wanted to breastfeed . . . and they said "Fine. Fine." And you stayed five days back then and the night before I'm ready to go home I could hear [my baby] screaming in the nursery. It was all the way down the hall but I knew it was her and I'd think, "What is the matter with her?" And the last night [I was in the hospital] some friend, who was a nurse, stopped in and she said, "Why are you taking that pill?" And I said "They gave it to me." And it was a dry up pill, to dry you up.[33]

Writing twelve years later, in 1972, a breastfeeding mother recounted in *Parents' Magazine* the story of her recent encounter with nursing staff. She had been the "only mother in the hospital breastfeeding her baby, and at least one nurse made it plain [to her] that [she] was upsetting her routine by requiring this special attention." "Nursing babies take longer to feed than bottle infants," she wrote, adding that mothers of breastfeeding babies require "more kind words of encouragement and instruction" than did their bottle-feeding peers.[34]

Many nurses wholeheartedly agreed with natural motherhood advocates that there were widespread systemic problems when it came to breastfeeding in hospitals. Among these, there were those who took it upon themselves to try to effect a change, even if it was by helping just one mother at a time. As one nurse wrote in to the *American Journal of Nursing* in 1947:

> I am on your side to fight "the battle of the bottle." What can we do to convince more mothers that nursing their babies is a satisfying experience? I am a graduate nurse and the mother of three healthy, happy youngsters; each one was breast fed for over four months. At the hospital, one of my room-mates was an intelligent enthusiastic primipara who wanted to nurse her baby. "Here is our chance," I thought and set out to do my bit. Those feeding hours were the high light of our day. Recently she said, "I have told my friends what a wonderful help you were in the hospital. You were like an audience at a football game, cheering from the sidelines for my best efforts."[35]

Nurse supporters also took aim at education, hoping to improve not only their professional standing, but to improve the ability of nurses to meet the demands of postpartum mothers. As one group of nurses wrote in 1948: "Among the many problems and changing concepts of maternal and child care which challenge us today, none is more deserving of thoughtful consideration than the preparation of nurses who will have such an important part in the care of mothers and babies in the future. Basic obstetric nursing courses have concentrated on teaching the student the technics of hospital care during labor, delivery, and the immediate puerperium."[36] The authors went on to cite a 1948 study which showed that 52 percent of the 1,300 state-accredited schools gave their students absolutely no formal experience working in antepartal or postpartal care.[37] Decrying these shortcomings of nursing education, they argued that in order to insure that students "receive a clear concept of total patient care, proportionate emphasis should be given to all phases of the maternity cycle."[38] Twenty years later, however, it seemed that these same problems continued to plague nursing education programs. As one nurse in 1968 described her educational background, she complained about the continuing lack of training when it came to breastfeeding. "We were told that [breastfeeding] was very good for both the mother and the baby," she said, but "this was all we learned!" This lack of practical education had had predictable effects on her competence in working with breastfeeding mothers. When confronted by them, she admitted, "I, along with many others, usually hid somewhere until all the breast-feeding was done."[39]

Maternal Demands: Nurses Respond

Despite the persistent lack of breastfeeding education in nurse training, accounts by mothers indicate that things did change, albeit slowly and sporadically. As one mother explained, her attempts to breastfeed her first child, born in the 1960s, were met with unhelpfulness on the part of the nurses. By 1971, however, when she went back to the same hospital to have her second baby she observed a very different atmosphere. She wrote, "Nurses were actually encouraging mothers to breast-feed their babies." "By this time," she added, "it seemed, everybody had decided that babies who weren't breast fed were at a disadvantage."[40] While there was not widespread uniformity or consistency to these changes, the fact that they were occurring at all was a significant departure from earlier decades. Nurses responded to mothers' interests in breastfeeding at the professional and institutional levels by taking advantage of small changes in hospital policies that allowed them more leeway when it came to interacting with their maternity patients.

The practice of "rooming-in" is a primary example of how policy and institutional changes could have a dramatic influence on nurses' interactions with breastfeeding mothers. Beginning in the 1940s, the practice of rooming-in began to appear in experimental bursts around the United States.[41] The irony of the rooming-in movement, of course, was that hospital birth had only relatively recently become the experience of a majority of American mothers—the year during which hospital births surpassed home births was 1938.[42] Prior to the early 1940s, most women were still having their babies in their homes. Irony aside, however, rooming-in was implemented as a response to mounting scientific evidence that suggested mothers and infants did better emotionally, psychologically, and physically when they remained in close proximity to one another. As rooming-in became part of the lay public's vocabulary, more and more women sought it out and the experience often went hand in hand with less medical intervention during birth and breastfeeding on demand.

Dr. Jackson's rooming-in project demonstrated that women preferred the experiences of having their babies with them throughout their stay and that doing so increased breastfeeding rates. Furthermore, and more importantly from a medical perspective, neither the mothers nor their babies had appeared to suffer an increased susceptibility to infection. Dr. Jackson, though a physician herself, saw nurses as the keys to successful implementation of rooming-in and breastfeeding. As a result of Jackson's interest in training nurses, many nurses who would play important roles as supporters of breastfeeding came out of the rooming-in program at Yale, including the head of the nursing program at Yale, Kate O. Hyder. Betty Ann Countryman, a graduate of the Yale School of Nursing, went on to became a prolific writer and advocate on the subject of breastfeeding. Jackson included Hyder in her inner circle at Yale, along with the heads of the divisions of Pediatrics, Obstetrics, and the dean of the nursing school. The important role that Hyder played in the rooming-in project is emphasized throughout Jackson's personal and public writings. Jackson consistently praised Hyder's efforts, citing her enthusiasm for breastfeeding and her willingness to train the rooming-in nurses directly.[43] Follow-up interviews with mothers further validated the importance of the nurses to the success of rooming-in and breastfeeding. As one mother wrote: "Being in rooming-in has been an extremely pleasant experience. The nurses all have very sweet personalities, which is extremely important in dealing with new mothers and their babies. They are helpful in every way. . . . Thanks again to rooming in and one of the most pleasant experiences of my life!"[44] Within the context of the emerging ideology of natural motherhood, women's desires for rooming-in situations often overlapped with their interest in breastfeeding. As Jackson observed, "the proximity of mother and newborn infant may increase the American mother's wish and ability to breast feed, both of which appear to be remarkably on the decline."[45] Though the most famous, Jackson's rooming-in clinic was not alone. Other rooming-in options sprung up in hospitals around the country. Wherever these options became available, nurses received specialized training and played integral roles to the successful operation of the program. Nurses who participated in these efforts learned to work with mothers in their efforts to achieve the birth and infant feeding experiences they sought. Where nurses and mothers worked together, outcomes and attitudes seemed to improve all around. As one nurse reflected:

> One of the most important advantages of this plan is the opportunity it affords to encourage breastfeeding. . . . Several mothers who were not able to nurse their children previously told us they had a more abundant supply of milk when nursing their babies by "demand feeding." We encouraged this practice, instructing the mother to allow the babies to nurse whenever the infant seemed hungry. . . . Several mothers who were not planning to nurse their babies at all did nurse them because of our plan, and did so quite adequately.[46]

Nurses who were involved in these pilot programs, like that discussed above, were often described as "wary at first" but quickly "converted" to the new way of doing things once they saw the happy results. As the aforementioned nurse happily observed about her experience with working in a rooming-in setup, "The babies cried very little; this was especially true of those [who were] breast-fed."[47]

Despite the appreciation of the benefits and possibilities surrounding rooming-in that many nurses articulated, many of these same women often balked at the idea of implementing rooming-in practices on a mass scale. As one nurse and mother of two explained, "I have enjoyed having my daughter with me and we have both progressed so nicely." Yet despite her own positive experiences, she did not believe it could work as "a general policy" because it caused "much more work for the nurses." She also worried that if implemented more widely, the practice could lead to problems of contagion and infection in mothers and newborns.[48] The scientific management of mothers and infants continued to appeal to nurses, even when they themselves had had positive experiences with rooming-in and breastfeeding. The rooming-in setup undoubtedly appealed to women because it fundamentally encouraged a natural motherhood approach to maternity through its affirmation of a woman's abilities to deliver her baby with minor intervention and to breastfeed on demand, without supplementation. At the same time, historians have suggested that rooming-in setups during this era may have simply been one more way to instill a belief in scientific expertise.[49] While on an individual level, rooming-in may have inspired women to connect to their instinctual maternal identities, widespread institutional changes seemed unlikely. The bottom line for many nurses ultimately seemed to be that whatever the benefits, breastfeeding mothers cost hospital staff additional time, energy, and resources in a context in which all of these were in short supply.

For every nurse who was excited about improving the experiences of mothers in the hospital, there were those who could not move past these kinds of logistical difficulties. One common concern with rooming-in, for example, arose due to the division of nursing labor. In recounting her experiences working in an experimental rooming-in setup, nurse Shirley Lundgren noted: "In the nursery was a group of nurses who had special preparation in the care of newborn babies while other registered and practical nurses were assigned to the care of the mothers. One of the first questions to arise was—who would care for the babies in the mothers' rooms?"[50] Nurses often recognized shortcomings like this when new hospital projects, such as rooming-in or demand feeding, arose because their positions were embedded within the technological systems of the hospital.

Wherever hospitals catered to the desires of breastfeeding mothers, nurses struggled to mesh the priorities of the hospital system with the newly imposed administrative priority of pleasing their patients. As a result, there were nurses who chafed against what they saw as the unrealistic demands that new mothers often made. As a group of nurses wrote in 1947, "Patients appear to judge the profession (of nursing) by the answer to just one question, 'is the nurse friendly and cheerful and always on the job?' If the answer is 'No,' then no amount of technical skill can change the patient's opinion." On the other hand, they observed, if the answer was "Yes," then "the nurse is an angel, the hospital is the best in the world, and nursing is a wonderful thing."[51] This conundrum, however, embodied the ideological difference in expectations that nurses and mothers brought to the breastfeeding experience. Mothers wanted, and in fact needed, more than anything to have an emotionally supportive, self-assured breastfeeding guide. Every mother was unique, every baby different. As La Leche League often articulated, what a nursing mother needed above all else was someone to help maintain her confidence while she and her baby adjusted to one another.[52]

When nurses did get involved in helping a mother breastfeed, their scientific training, coupled with what was often their own lack of experience with breastfeeding, often proved harmful for breastfeeding mothers. Despite La Leche League's collective experience to the contrary, it was common practice among nursing staff to disinfect and clean women's nipples before and after breastfeeding. Not only did women describe this as being less than relaxing (and thus, inhibiting their milk let-down reflex), it also could easily lead to overdrying and cracking of the nipples.[53] In addition to being painful, dry and cracked nipples were preconditions for more serious problems, such as infection (mastitis), or even an abscess. Furthermore, even in cases when nurses followed a mother's wishes and brought her infant for breastfeeding, rigid feeding schedules in the hospital meant that breastfed infants were typically kept to the same four-hour feeding intervals as their bottle-fed peers.[54] One nurse summed up the lack of experience and expertise that plagued mothers' hospital experiences when she noted that "young women unacquainted with breast feeding often became the nurses who staffed maternity divisions."[55] She added that when nurses arrived on the maternity floor, they had perhaps "been taught breast anatomy and physiology," but could rarely "apply this knowledge to assisting mother and infant." The scientific management of breastfeeding, argued natural motherhood advocates, was practically impossible.[56]

Cultivating Nursing Support and Professional Autonomy

Despite these difficulties, breastfeeding continued to gain supporters in the nursing community, particularly among the more extensively (often college) educated nurses. Those at the top of their field in nursing were most likely to support programs like rooming-in projects that supported breastfeeding. These reform-minded nurses led the movement to humanize maternity nursing care from within the profession itself and urged their fellow nurses to take an interest in breastfeeding. Nurses like Yale's Kate O. Hyder and Betty Ann Countryman seized on the evidence available to them from the social sciences, pediatrics, and progressive medical institutions as well as key physicians (such as Edith B. Jackson, or in Countryman's case, her husband, a doctor of psychiatry) to establish postpartum care and breastfeeding as an area of nursing expertise.[57]

From the perspective of these nursing leaders, breastfeeding offered nurses the opportunity for greater professional authority. Breastfeeding in the hospital did not typically require a physician's oversight, nor did it even, in many cases, pique most doctors' interest. If women breastfed and nurses were the experts who helped them, then nurses might increasingly claim jurisdiction over the postpartum care of the mother and infant. The hands-on, personalized nature of breastfeeding defied centralized oversight and routine and therefore could not be easily written into standing orders in the same way that formula feeding could. Nurses who embraced the professional possibilities that breastfeeding offered them attempted to convince their colleagues that breastfeeding was not just about demanding, self-righteous mothers, but that it could also be about claiming authority and power over the care of patients in the hospital. In doing so, nurse breastfeeding advocates and mothers faced decades of the entrenched ideology of scientific motherhood, which placed greater value on all things scientific and technological. Formula feeding was believed to be the more scientific way to feed an infant and hence held greater value and prestige for many nurses educated in the traditions of scientific motherhood. In the middle of this ideological shift in nursing sat La Leche League, whose goals had long included gaining support from the medical and nursing professions.

Nurses became aware of the League through a variety of means, including becoming mothers themselves who were in need of breastfeeding help. Some encountered patients who were La Leche League members, while still others learned about the League through fellow nurses and physicians. Many more learned about the League through informal visits from its members as they made their way around area hospitals, spreading the word about their group's presence and mission. Though the League began outside of Chicago, it quickly spread. In 1964, the League had made its way to Jersey City, New Jersey, carving a path for breastfeeding among the formidable and historically entrenched hospitals of the Northeast. One nurse, Josephine Iorio, heard about the Jersey City group and soon found that the League could be a great resource for nurses, a discovery she recounted in a letter to the *American Journal of Nursing*. "I give new mothers the names of La Leche members," she pointed out "so that, if at home they have any problems or questions, they can call for help."[58] But the League was not only useful because it provided nurses with a quick and easy referral network; it also served as an educational resource for the nurses, themselves. "Meetings are open to nurses and those of us who have attended have learned better how to help patients," she wrote.[59] "For instance," she explained "at one meeting a number of mothers remarked that nurses are very inconsistent in what they tell mothers about how long to keep the baby at breast for a feeding, whether to use one breast or two, and so on." The friendly atmosphere and the open discussion had positive results. "Nurses who were at the meeting discussed this comment at a hospital conference," she recounted, adding that the hospital staff were now "clear and consistent in advising" when it came to breastfeeding.[60]

The League's involvement with nurses was an important aspect of their larger effort to increase support for women who wanted to breastfeed. In 1972 they published a pamphlet specifically for nurses, highlighting the importance of the nurse's role in the mother's early days in the hospital. "Your *praise* and *reassurance* are vital," the pamphlet emphasized. "Having a nurse's knowledgeable *support*, *encouragement*, *understanding*, and *patience* while she is in the hospital will help the nursing mother to leave the hospital relaxed and secure in her relationship with her new baby."[61] This was important, they explained, because "appropriate advice, encouragement, and support during the days in the hospital lay the foundation for an easy, natural beginning of the breastfeeding relationship and minimize maternal anxieties and other difficulties in the weeks ahead."[62] The League emphasized the good nurse as a caring and soothing presence for the anxious new mother struggling to get her baby to latch on, while she coped with sore nipples and waited for her milk to come in. Of utmost importance in those early days in the hospital, the League emphasized, was the mother's need for the presence and attention of the nurse. The League wrote, "Please *do* stay with the mother as long as she needs you during the first few feedings. She'll need your emotional support as well as information and guidance about many things."[63] The League stressed the nurse's importance as an emotional support for the mother, even while it de-emphasized any scientific or technological interventions into the breastfeeding process. The best medicine in this case, the League seemed to say, was no medicine at all. The League's philosophy toward nurses emphasized the important role they might play as extensions of a broader female network of maternal expertise. "Above all, we encourage you to take a warm, positive approach to *all* the new mothers who come to you," the League wrote.[64] All mothers, they added, "need your support, reassurance and encouragement so that they will grow in mothering and nurturing their babies."[65]

Evidence abounded from follow-ups with mothers' and women's letters to journals and magazines that this was, in fact, sound advice. As one mother summed it up, "Just [the nurse] being there with a comforting word and some practical know-how was so important to me. I really needed her. She gave Jimmy and me a good start."[66] In accounts such as this one, mothers rarely, if ever, mentioned the nurse for her *scientific* expertise on breastfeeding. The nurses whom breastfeeding mothers singled out for praise and recognition were those who had been willing and able to simply "be there" with them.

La Leche League did not operate as an outside organization in its endeavors to increase nurse's awareness of breastfeeding. Leaders within the field of nursing often wrote on their own in professional and popular venues, encouraging their colleagues to embrace a more compassionate approach to postpartum care. As one nurse put it, "The woman who wants to nurse . . . needs all of the support and encouragement which she can get, starting with the nursery and postpartum nurses."[67] For these nurses, "support" was, in fact, seen as distinct from a nurse's usual technical responsibilities. As breastfeeding advocate and nurse, Josephine Iorio explained in 1963, "The important thing . . . is to show the mother your very keen interest in helping her and her baby to make a good start." In discussing her nurse-training program, Iorio stressed that her nurses were urged to "stay with the new mother until her baby is nursing properly." Doing so, she told them, would not only keep the baby on breast milk, it would render the mother "extremely grateful."[68]

Motherhood Changes Everything: Breastfeeding and the Nurse-Mother

Despite the efforts of La Leche League and nursing leaders to improve nursing care, nurses as a professional group were slow in heeding their advice. As a 1983 position statement from the Massachusetts Nurses Association illustrates, the failure of nurses to uniformly and consistently provide emotional support to breastfeeding mothers persisted well into the final decades of the twentieth century. They wrote: "The Massachusetts Nurses Association believes nurses have a great impact on infant feeding practices and must therefore be aware of their power and influence. They must examine their own beliefs about infant feeding practices to prevent personal bias from hindering the education and direct care which they provide to parents."[69] Evidence has suggested, in fact, that the most important factor in being a successful breastfeeding coach is personal experience.[70] Due to slow and sporadic changes in nursing education and institutional policies, the majority of nurses (and physicians) throughout the second half of the century had little formal education on breastfeeding. As a consequence, nurses often drew upon their own personal experiences whenever possible when it came to advising mothers on the practice.

The 1972 La Leche League booklet "How the Maternity Nurse Can Help the Breastfeeding Mother" was authored by a group of maternity nurses who were also mothers themselves. Although the formation of a personal infant feeding philosophy is and always has been a complicated process, for nurses there really were two divergent paths—either one's own experiences with infant feeding urged them to support breastfeeding for mothers in the hospital, or it led them to be dismissive or discouraging of the practice. The author of the La Leche League pamphlet's preface, mother, nurse, and League supporter Betty Ann Countryman emphasized the unique expertise of breastfeeding nurse-mothers:

> [The] child health nurses who helped to prepare this booklet—experienced breastfeeding mothers all—know first-hand the many problems which may arise and with which the nurse, as well as the nursing couple, may need help. We have experienced our own deep sense of gratitude toward nurses who have been patient, understanding, and supportive, and in turn, we—as nurses and mothers—have felt the glow of inner satisfaction which has followed a successful attempt to help the inexperienced mother grow in self-confidence and learn the joys of mothering and nursing her baby.[71]

It was not uncommon for nurses to experience the birth and postpartum feeding of their own infants as a revelation, a transformative experience that could have dramatic consequences for how they approached their work. As nurse Ruth Owen wrote in 1951, the experience of motherhood changed her in just such a fashion. "You will think it nothing out of the ordinary when I say 'I have a baby!' But for me having that baby was a revelation. I really saw myself 'as others see me.' If ever again I practice nursing I know *my own experience will guide me* [emphasis mine] and that every woman in labor I care for will benefit."[72] Nurses not only experienced shifts in perspective once their children were born, they could just as importantly also find a new source of expertise and professional authority.

Nurse-mothers of this sort, such a Margaret O'Keefe, carved out a rewarding career by combining experiential knowledge about breastfeeding with their professional roles. O'Keefe first published as a "mother, and a nurse" in the *Ladies' Home Journal* in 1962.[73] One year later, she was publishing as an expert on the subject of breastfeeding in the *American Journal of Nursing*. In her 1963 article, O'Keefe wrote that since nursing her first child, she had found it impossible to ignore the "very real cry of the new nursing mother for constant support and encouragement from an experienced, competent helper."[74] Citing her own breastfeeding experience as a turning point in her life, O'Keefe reported that she had since "personally helped more than 200 women to breast feed without one real failure."[75] Her advice to hospital administrators and physicians followed accordingly: take advantage of nurses who have successfully breastfed their own children. "A registered nurse who has successfully breast fed her own babies or who is enthusiastic in her attitude," she wrote, is better able to provide calm and consistent encouragement to new mothers "than most overworked doctors who could then direct their time to the medical control and treatment which they alone can provide."[76] O'Keefe crafted her personal experience as a breastfeeding mother, alongside her professional training as a nurse, to carve out an authoritative niche within hospital maternity care—a niche that was notably outside of the domain of the busy medical doctor and that could not be efficiently filled by someone without breastfeeding experience. O'Keefe, and others like her, utilized their dual roles as mothers and as nurses to advocate for breastfeeding and to create a more fulfilling career for themselves within the context of the overbearing routinization of the hospital maternity ward.

Other nurses who were less successful in their personal breastfeeding endeavors took slightly different approaches in their advocacy efforts. Lorraine Weszely, a nurse who wrote in to the *AJN* in 1968, complained that in spite of her training as a nurse she had suffered through four weeks of frustrated breastfeeding before giving up and switching to formula. She found, ironically, that her nurses at the hospital were no more knowledgeable on the matter than she was, and that they had little time or concern for her desire to breastfeed. "Some nurses would come into the room, slap the baby around until it woke up, literally stuff the baby's mouth to my breast, and leave," she recounted. With her second child, she took it upon herself to read everything she could and found she was more prepared to succeed on her own, but the experiences left her changed and wondering at the lack of education for nurses on the practice of breastfeeding: "In my opinion, the birth of a child and its subsequent care are the most important things in life for a mother. If a mother decides to breast-feed her child, it seems only right that she be able to turn to the medical profession for help. I feel that something as personal as breast-feeding can be handled best by a nurse, instead of by a neighbor or friend."[77] Motherhood, it seemed, put the nursing profession into perspective for many nurse-mothers, who returned from their postpartum experiences with fresh eyes on the subject of breastfeeding.

After eight years as an obstetric nurse, Jean Cotterman experienced a dark sense of failure when she found herself failing at breastfeeding her fourth child, just as she had with her first three. She humbly recounted in 1966 how it took a group of "lay women to rescue me" (she was referring to La Leche League). Cotterman wrote about her experiences in *Marriage: The Magazine of Catholic Family Living* and prefaced her article on the hurdles and victories of breastfeeding by recounting how motherhood and her involvement with La Leche League had changed how she thought about her career. At La Leche League meetings, "unshielded by clinical surroundings," she wrote, "I often felt secretly embarrassed for my profession." She eventually came to think more critically about her role as a nurse, reflecting: "I often wondered how many mothers I had helped to start toward dissatisfaction."[78] Haunted by these past encounters with breastfeeding mothers, Cotterman adopted a humbled and more sympathetic perspective as she moved on in her career. Assisting mothers with helpful breastfeeding advice, Cotterman carried the holistic model of the mother-infant breastfeeding relationship with her into the maternity ward. In doing so, Cotterman carved out a more fulfilling career for herself, born from her experiences as both a breastfeeding mother and a nurse. It is unclear, however, how many women like Cotterman actually returned to nursing work. While O'Keefe returned to some kind of hands-on work with mothers, it is not clear, even in her case, to what extent she remained employed as a nurse. Many women who could afford to embrace breastfeeding to such a degree were those who were financially able to leave their work behind them while they pursued motherhood as a vocation, even if it was only for a period of a few years.

The transformations in nurses like O'Keefe and Cotterman, though important, are not meant to be representative of the experiences of all nurse-mothers. Many of the nurses who were so often not the ones writing articles for the *AJN* or any publication, for that matter, were mothers of young children who worked part-time or even full-time in order to provide for their families. Nurse-mothers in this position often had a much different perspective on the importance and sentimentality of breastfeeding. As nurse Karen Martin wrote in her letter to the *American Journal of Nursing* in 1972, "Breast feeding isn't for everyone."[79] It certainly had not been for her. She recounted her feelings of resentment at "having to defend my choosing to bottle-feed" and complained that "as an obstetrical nurse" she had been "frustrated by unsuccessful breastfeeders who are breast-feeding simply because they felt pressured to do so."[80] Martin voiced her belief that nurses, regardless of their personal feelings, should be held responsible for letting mothers choose the method of infant feeding they wanted and that alerting mothers to the option of bottle-feeding was an important part of being an obstetric nurse.

Ideological Differences

Mothers' accounts confirm the presence of a cacophony of voices in the maternity ward when it came to infant feeding support. As one mother explained in a 1972 account to the Boston Association for Childbirth Education, "When requesting help with getting baby started breastfeeding, everyone on the staff had different ideas. I was very perplexed. Now that I'm home all is well."[81] It is a difficult endeavor to piece together the various perspectives that nurses brought with them to their work with mothers. The familiar problem of the written record is one large source of this difficulty; the vast majority of nurses who took the time to write articles or letters for publication were representative of the middle class, a segment that, while vocal, did not represent the bulk of nursing practice during the period in question. Occasional letters do provide insight, but a particularly useful source is a 1974 study on the views of nurses and La Leche League mothers.

In her PhD work in sociology at Northern Illinois University, Kathleen Knafl analyzed data from over 940 responses (out of 1,000) from La Leche League mothers and nurses. Her findings add further support to the historical evidence that the typical encounter between a nurse and a breastfeeding mother was often plagued by fundamentally different ideological views on motherhood and infant feeding. Knafl described the typical nurse respondent as being unenthusiastic about the benefits of breastfeeding over bottle-feeding, believed it to be "troublesome," and particularly disliked La Leche League members whom they often viewed as "fanatics." From the nineteen survey responses that she analyzed in depth for her 1972 *AJN* article, Knafl argued that the majority of the nurses believed that breastfeeding was "more convenient" for mothers but that it was "disruptive of hospital routines, when they had other mothers and duties to consider."[82]

Knafl also reported that her nurse respondents were more likely than the League mothers to articulate a belief that breastfeeding and bottle-feeding were equally good, and they were quick to point out that working mothers had a more difficult time breastfeeding for logistical reasons. Knafl's nurses overall found it difficult to justify what they saw as breastfeeding fanaticism on the part of many breastfeeding mothers. For them, even as late as 1974, bottle-feeding was literally just as good as breastfeeding, and scientific motherhood remained the dominant ideology to which they subscribed. The nurse respondents openly admitted to profiling and judging their patients based on their stated desires to breastfeed or their affiliation with La Leche League. Knafl found that "if a nurse identified a mother as a member of La Leche League and defined her requests as demanding, the possibility for conflict between them increased." Nurses overwhelmingly believed that formula feedings of infants in the nursery were necessary and that mothers' requests to the contrary could be and often were ignored. As one nurse put it, breastfeeding mothers could be "a bit paranoid about it."[83]

Knafl's report provides insight into the persistent conflicts and stubborn ideological differences that continued well into the 1970s, a period during which breastfeeding rates would begin to turn around. As larger numbers of mothers became more interested in natural motherhood and breastfeeding, nurses often grew impatient and annoyed by these demanding patients. Knafl's nurses reported making and acting on professional judgments about a woman's "suitability" for breastfeeding, regardless of a mother's stated desires.[84] Furthermore, those mothers whom nurses judged to be too nervous or anxious were those most likely to have their breastfeeding efforts ignored or hindered. Knafl's report suggests that nurses, on the cusp of breastfeeding's resurgence, struggled against a growing tide of expectations that advocates of natural motherhood brought to the hospital experience. In 1974 the nurse respondents were adamant about their continued belief in the necessity of a scientifically managed mother-infant pair. Through this position, they articulated a persistent regimen of scientific motherhood that focused on strict schedules, medical management, and formula feeding even as leaders in the fields of nursing, pediatrics, and a growing community of lactation activists were beginning to shift to a position in support of breastfeeding. At the same time, natural motherhood advocates, empowered by renewed scientific support of the benefits of breastfeeding, continued to maintain that experience-based support needed to be integrated into the hospital setting. With their stake in scientific motherhood, nurses often found themselves in opposition to the return of breastfeeding.

Making the Nurse Visible Again

Understanding the perspectives that nurses and mothers brought with them to the hospital encounter complicates the historical narrative of breastfeeding history by challenging the primary importance of the physician and formula companies in women's decisions, experiences, and success in breastfeeding. Nurses' experiences and beliefs surrounding breastfeeding throughout the decades following World War II shaped their responses to mothers' demands for breastfeeding in the hospital through a variety of approaches, bounded as they were by the institutional restrictions placed on their autonomy by the hospital itself. The nurse's desire and ability to effect changes in breastfeeding assistance in the hospital were rooted in her relationship with her patients, a relationship profoundly shaped by the tensions of competing constructs of twentieth-century motherhood and by conflicting ideas about the place of medical science and technology in infant feeding practice.

The interactions between mothers and nurses were highly gendered, based on their sometimes shared—but often conflicting—assumptions, expectations, and experiences (or lack thereof) of motherhood. The nurse, a trained professional and avid believer in scientific medicine, had been, for decades, an evangelist for the ideology of scientific motherhood. By the late 1940s, however, this construct of womanhood was being challenged by the emergence of an alternative maternal ideology, that of natural motherhood. Nurses who had experienced successful breastfeeding for themselves occasionally found it easier to evade the pressures of the hospital and of physician oversight in order to connect in a meaningful way with mothers. In doing so, these nurse-mothers forged an alternative career path for themselves through their dual identity and their embrace of natural motherhood and breastfeeding. The nurse-mother was an expert with a tacit knowledge based on maternal experience and instinct rather than scientific education alone. These women often were able to achieve a specific aura of authority over breastfeeding knowledge, which allowed them to connect with mothers while still maintaining their professional identity as a scientifically trained health care professional. As women and as health care professionals, nurses held a unique social position as the breastfeeding movement gathered momentum by the early 1970s. Poised to usher in an era of medical support for breastfeeding mothers, nursing as a profession continued to struggle over the tensions raised by the ideology of natural motherhood.

Far more nurses throughout the postwar decades did not identify with their breastfeeding patients. These breastfeeding detractors (either in principle or in practice) often described mothers' desires to breastfeed as frivolous, overly demanding, and even fanatical. Rather than identifying with their patients as fellow women and/or mothers, many nurses felt distanced from their patients precisely because they wanted to breastfeed. Nurses who did not agree with or support a mother's choice to breastfeed saw themselves as scientific laborers forced to do less-than-scientific work, and they disapproved of the rhetoric and philosophy surrounding natural motherhood. Fully entrenched within the scientific and technological system of the hospital, these nurses actively struggled against women's desires to breastfeed because it conflicted with their idea of an appropriate, scientific model of motherhood. Many more nurses simply failed to see a way around what in their view was the ideological disconnect between the realities of postwar medical care and the desires of mothers to embrace an ideology of natural motherhood, particularly when it came to breastfeeding. As the back-to-the-breast movement grew, nurses would not find themselves unified on the forefront of an emerging area of expertise. Instead, the profession as a whole would remain committed to the ideology of scientific motherhood and the rigid protocols of scientific maternal and infant care. As more and more mothers demanded breastfeeding support, however, a new group of experts would rise to prominence first in the form of La Leche League counselors and, later, as an independent profession of lactation consultants. In the remaining chapters, the widespread shift back to the breast that became fully realized in the 1970s is explored in detail. Despite its faltering start, breastfeeding would emerge again as a central, if not highly contested, component of maternal identity by the close of the century. The question of how to integrate breastfeeding, with its connection to the midcentury's ideology of natural motherhood, loomed large in these years as lay groups and professionals alike grappled with its rebirth in light of modern feminism, environmentalism, and public health efforts.

CHAPTER FIVE

Our Bodies, Our Nature

Breastfeeding, the Environment, and Feminism

Born in 1940 in Brooklyn, New York, Patty and her family later moved to Yonkers when she was five, "where the entire panoply of nature was still vibrantly intact." When she was nine years old, her family moved again, this time to south Florida. Compared to the forests and meadows of 1940s Yonkers, Florida appeared to her a rural wilderness, unlike anything she had experienced before. "I became an avid birdwatcher, a collector of fossils," she recalled. "I studied wildflowers, insects, ecology. When I herded cattle in the summers, we often helped with the difficult births of foals. . . . I lived close to nature in every sense." Her early love of nature and her exposure to the natural world, particularly the animal world, left her with a lasting feeling that doing things "naturally" was the best way for her to live. This attitude transferred seamlessly into her identity as a mother when, years later, she chose to breastfeed her children. "I loved animals and all natural processes and I know this shaped my feelings about natural childbirth and breastfeeding later," she recounted.[1]

Growing up in the World War II and postwar years, Patty also experienced what she described as a considerable degree of alienation from the mainstream culture of America at the time, with its emphasis on technological fixes, the consumption of mass-produced goods, and proscribed gender roles. A story she shared about her love, as a small child, of oranges over the manufactured sweetness of store-bought ice cream helped emphasize her long-held disdain for a popular consumer culture she did not support. In high school, a favorite teacher included her as the lone female in his all-male group of select students who received invitations to dinners at his home. This encouragement, she explained, helped her feel that she could transcend what she felt was the hoped-for role of the American woman-as-housewife that she felt dominated the youth culture surrounding her. Finding refuge in "nature" as opposed to the society of postwar America, Patty linked her distaste for the era's gender roles with her affinity for what she experienced as the freedoms allowed by nonhuman nature in her life narrative. Her love for the outdoors, for education, for the natural world, and her dislike of cultural gender norms and the "superficiality" of postwar American life led her to seek experiences in Europe as a young woman in the early 1960s. She was most impressed by the time she spent in Italy, particularly when it came to motherhood. She recalled:

> I found things were far more natural in Europe. . . . The Italian women blew me away, I loved it there. . . . They would tuck their big skirts up into their belts and they would go into the streams and bang the sheets and the clothing against the rocks . . . and then they would spread all these things out, no soap, on the hillsides to dry. . . . So you had all the babies, children, running around. Babies at the breast, mothers taking turns watching everyone else's babies and children until all the women had a chance to do their laundry. What a fine system—both social and pragmatic![2]

In 1969, Patty gave birth to a son in a Philadelphia-area hospital. "I knew that I wanted to breastfeed—well, because it delivers all the antibodies, it's not just cute! It's just much better for allergic stuff later and it's just the best thing you could do. . . . I mean, nothing in a bottle or in a can or something could do it, it's crazy, so I knew that's what I wanted to do."[3]

Despite her own convictions, breastfeeding remained an alternative practice for the majority of mainstream Americans through the early 1970s, with fewer than 30 percent of mothers breastfeeding while in the hospital. Even with her decision firmly in place, then, it is not surprising that during her hospital stay Patty experienced a lack of support and interest from doctors and nurses in her desire to breastfeed.[4] She "struggled through" on her own, however, and did the same with her second son. By the time she had her third child, Patty was able to connect with a local breastfeeding support group that was forming called Nursing Mothers.[5] Once Patty connected with Nursing Mothers, she found a community of women, many of whom (though not all) shared her overlapping values of motherhood, nature, and family. "I'm a big walker . . . whoever's a walker is usually a breastfeeder . . . if you're a birdwatcher you probably breastfeed your baby," she recounted.[6] In addition to drawing direct connections between the animal and human worlds, Patty also emphasized the connections between breastfeeding and doing things more simply, and with less reliance on complex technologies. In addition to the embrace of breastfeeding and the nurturing function of the female breast, Nursing Mothers also helped its members learn about and obtain simple technologies that supported their approach to a more "natural" lifestyle. "We discussed people all over the world carrying their babies on them. . . . I didn't put my baby down [and] I think they're much more grounded." Patty and other mothers in her circle utilized baby slings that were made available for purchase through the group itself. In keeping with their concern about "the natural," Patty recalled, she and many of her peers belonged to some of the earliest food co-ops in the Philadelphia area: "I had always been very much up on nutrition and I never had a hamburger, one of those McDonald's in my life!" Not surprisingly, then, Patty was (and still is) thrilled with a small, hand-powered device that she discovered in the 1970s, while in Nursing Mothers: "It's a little hand mill." Me: "So it's not electric?" Patty: "No! . . . Whatever's at the table you throw it in, you grind it and you feed [it to the baby]! . . . They're like poisoning the children in this country [with processed food]!"

Functioning in much the same way as La Leche League, local groups like Nursing Mothers offered women an opportunity to connect over their shared belief in breastfeeding but it also provided a network of motherhood expertise built upon experience, support, and the understanding that breastfeeding issues could transcend infant feeding. "Everybody supported everyone else and sometimes somebody was having trouble with their husband, he wasn't [being] supportive [or] he didn't want you to breastfeed and [we] would say 'yeah, shit on him!' You know? . . . so we empowered each other. So it became far more than just a cute, sociable little meeting. It was really a wonderful support group." Initially manifest through breastfeeding, Patty's embrace of natural motherhood continued to influence her life even as her two children grew. "I never bought the latest [thing]. . . . The kids loved the outside, their favorite thing was a pail of water and dirt. . . . They loved it, and they made mud pies and . . . that was wonderful." For Patty and a growing number of other mothers throughout the 1960s and 1970s, the discovery of "the natural" became an important and central component in their approach to motherhood.

This chapter explores the intersections between the concept of nature that Patty and many other breastfeeding women subscribed to in the 1960s and 1970s within the breastfeeding movement. The interaction between ideas about nature and gender in this period both fueled and complicated the move back to the breast throughout the 1960s, 1970s, and 1980s. It is no coincidence that renewed interest in breastfeeding emerged alongside the early rumblings of the modern environmental movement and prompted a debate within feminist circles that to a significant extent persists to this day. The appeal of "the natural" helped to drive mothers' interests back to the breast, but once there, more and more women came to question the compatibility of the ideology of natural motherhood with modern feminist ideals.[7] Issues regarding women's labor, child care, and reproductive health rights more generally helped fracture the breastfeeding community by the 1980s. The splintering of the movement would have lasting implications, including the emergence of a rhetoric of "choice" that has often obscured or downplayed the structural realities of mothers' lives. Mothers struggled throughout this period to fit breastfeeding into their increasingly diverse lives, eschewing aspects of the ideology of natural motherhood in favor of a more modern ideology of choice and individual achievement.[8]

In Search of Modern Mother Nature

In the Western world, breastfeeding, a decidedly female activity, has been explicitly marked as "natural" since male scholars in the age of the scientific revolution began carving the world into separate poles of machine/male and nature/female.[9] What exactly it has meant to call something "natural" however has varied, and it is the argument of this book that the link between nature and breastfeeding assumed heightened importance over the second half of the twentieth century. In analyzing how breastfeeding came to be seen as natural in the mid-twentieth century, it helps to look to the history of one of America's "most natural" foods: cow's milk. The unique yet relevant history of America's infatuation with drinking the milk of cows offers a useful framework for thinking about the changes in ideas about the connections between milk, bodies, and "the natural."

In *Nature's Perfect Food: How Milk Became America's Drink*, sociologist E. Melanie DuPuis argues that American ideas about nature, particularly the healthfulness of "the country" in opposition to "the city," helped dairy farmers and milk marketers over the course of the early twentieth century sell Americans on the idea that cow's milk was a uniquely complete, healthful, and necessary part of the balanced diet.[10] This became particularly true for women of childbearing age and children as the century wore on. By the 1950s, children especially served as the idealized consumers portrayed in advertisement campaigns for cow's milk, a product that was increasingly produced via industrialized agriculture rather than on idyllic and pastoral family farms. In a particularly clever analysis, DuPuis shows how farmers and marketers helped shift the public's gaze away from the disintegration of the "natural" in the modern industrialized dairy toward the consumption of milk by "natural" and healthy mothers and children. By shifting the focus of their marketing from the symbol of the dairy maid (the pure embodiment of the American countryside) to the symbol of the mother and child, these advertisements also reveal a broad change in how Americans understood their relationship to nature. Rather than being something embedded in a particular landscape, the nature reflected in changing constructions of cow's milk suggested that nature could be consumed and even embodied. As DuPuis points out, "In this new view of perfection, the nature involved in the production of the product also changed. Rather than the nurturing feminine countryside, nature became 'matter,' a fuel used for the sustenance and growth of human machines."[11] DuPuis's analysis highlights a subtle yet wide ranging shift in American ideas about nature as cow's milk itself became valuable as a material connection to "the natural" even as fewer and fewer people participated in, knew about, or even saw "the country" it purportedly represented.

By the 1950s, cow's milk had become a cornerstone of the American diet, signifying a connection to "the natural" at the same time that fewer and fewer infants received milk from their own mothers. Mother's milk itself, however, was poised to undergo a similar shift in terms of its relationship to evolving understandings of "the natural." By the end of World War II, aided by growing concerns about contamination of the environment and the industrialized food supply, more and more Americans looked to mother's milk as a reservoir of nature and purity that remained untouched by modern technologies. Throughout the postwar decades, Americans learned of nuclear fallout, food contamination incidents, and the harmful effects of pesticides on their own backyards, while the rising specters of cancers and childhood poisoning loomed ever larger. At the same time, the connections that psy-entists had helped to build between the human and animal worlds continued to destabilize mechanistic notions of a human body that existed separate from nature. In this context, natural motherhood ideology came to rest on the idea that the purest and most natural milk for human consumption could only be *mother's* milk. The implications of this were profoundly important for the creation of a new orientation toward breastfeeding in which women could in fact embody "the natural" through the biological acts of motherhood, particularly breastfeeding.

Getting Back to Nature

Patty's relationship with the natural world may have been unique to her in some respects, but an interest in getting "back to nature" was something shared by many women, and men, in her generation. A 1971 letter published in *La Leche League News* emphasized the ideological symmetry between the embrace of breastfeeding and a broader ecological mind-set. The author wrote:

> One of the concerns of today is the amount of trash we human beings throw out every day. Well, now just think of all the condensed milk cans, prepared formula cans, disposable formula sacks, canned juice cans, baby cereal boxes, and baby food jars that are never opened by LLL mothers and are never added to that trash. Another big item that we're told is important is getting back to natural products. They are emphasizing natural foods, natural soaps, and natural ingredients of all kinds. What could be more natural for human infants than human milk?[12]

As this suggests, breastfeeding seemed an easy fit for the emergence of a more general cultural interest in "the natural." It was this growth in ecological awareness and a longing for a connection to nature that led many women of Patty's era to embrace the ideology of natural motherhood when it came to things like breastfeeding, being more self-sufficient, and to some extent rejecting the postwar belief in consumerism and technology. Nowhere was this perspective more apparent than among those who participated in the "back to the land" movement, in which thousands of primarily middle-class families packed up their station wagons and headed for the woods to see if they could live a better life by being closer to nature.

Described as a reaction against the manicured, paved, and highly managed world of postwar suburbia, the move back to the land combined a quest for a natural life with attempts at forming successful and happy family and community structures. In the 1970s, somewhere between 750,000 and 1,000,000 individuals lived in back-to-the-land communes, a number that author and back-to-the-lander Eleanor Agnew has suggested would be much higher if the independent homesteaders and small farmers of the period were also counted.[13] By the early 1980s, as author Jeffrey Jacob has suggested, as many as twenty million Americans were sympathetic to the idea of the pursuit of a "simpler life."[14] Against this changing backdrop of American cultural values, natural motherhood through breastfeeding appealed to this new generation of mothers precisely because it emphasized "nature" as a guide for living.

While perhaps only one million Americans ever took the plunge into true homesteading, many more sympathized with the search for a more "natural" existence, free from what they believed were the harmful effects of big science and modern technology and industry. With the 1962 publication of Rachel Carson's *Silent Spring*, Americans learned of the dire consequences of the twentieth century's ascent into technological modernity. The pesticides and other chemicals that silenced Carson's birds were also making their way, on a cellular level, into the bodies and lives of humans. A series of articles published in *Nature* in the 1960s discussed the presence of cesium-137, a radioactive isotope produced by nuclear reactions, in breastfed babies in Europe. By 1970, scientists were supporting the argument that human breast milk seemed to have measurably higher concentrations of these radioactive components than processed formula.[15] These findings merely compounded parallel concerns over the presence of pesticides and other harmful substances in food, particularly human milk. As Patty's story suggests, however, the natural movements of the 1960s and 1970s had important roots in the 1930s, 1940s, and 1950s. The origins of the postwar movement back to nature, like many of the other ingredients of natural motherhood ideology, can be found in the interwar period.

By the late 1930s, the work of J. I. Rodale could be found on magazine and book stands thanks to the success of his publishing body, the Rodale Press. An early and avid evangelist for the connections between natural foods, soil health, and human health, Rodale expanded his ideas beyond the publishing world into the operation of his own organic farm in Emmaus, Pennsylvania, in 1940. Enthralled by the organic farming work of British agriculturist Sir Albert Howard, Rodale translated his enthusiasm for composting into an ideological empire for organic farming and "natural" living. By May 1942, Rodale was publishing a popular magazine on implementing these principles called *Organic Farming and Gardening* magazine (changed to *Organic Gardening* in January 1943), and by the end of the decade he had cultivated tens of thousands of subscribers from across the nation. Rodale's 1945 book, *Pay Dirt*, which offered the latest in research and testimonials on organic soil science, sold fifty thousand copies and went through three editions.[16] When Rodale launched *Prevention*, a "magazine devoted to the conservation of health," in 1950, he attracted fifty thousand subscribers before the first issue was released in June of that year.[17] It was in *Prevention* that many Americans first learned about the potential dangers of dichlorodiphenyltrichloroethane, or DDT, in 1952. That Rodale influenced American culture, and agriculture, is indisputable, but Rodale's organic health movement overlapped in significant ways with other early "natural" movements that focused on environmental contaminants and the connections between nature, human bodies, and infant and child development. In fact, copies of his *Organic Gardening* magazine and other publications can be found tucked away in the files of La Leche League's founders.[18]

By the mid-1950s, Rodale was a notable part of a growing contingent of Americans seeking healthier lives through more natural routes. Maternal health reform organizations like La Leche League were hard at work spreading the message of natural motherhood through breastfeeding, and, with the formation of the National Committee for a Sane Nuclear Policy (SANE) in 1957, concerned citizens came together to educate the public on the connections between environmental contamination, armed conflict, and human health. Some of the earliest responses to environmental contamination, in fact, emerged in light of fears over radiation fallout from nuclear weapons testing and the effect this fallout could have on developing infants and children. At the center of these popular reform movements in maternal and environmental health was a group of women with common traits—they tended to be white, middle-class, married mothers, often with college educations, who lived in the suburbs of large metropolises throughout the Great Lakes and Northeast regions.[19] These women, as one historian has remarked, "sought to reassert their role as protectors of the home and children," a domestic role that came under attack in the postwar years.[20] For these environmentally concerned denizens of the domestic sphere, news that radioactive fallout, DDT, and polybrominated biphenyls (PBBs), among other things, were corrupting the food chain was a figurative call to arms.

SANE's members may have sought above all to end nuclear weapons development, but as other historians have pointed out, many of their public awareness campaigns explicitly targeted the concerns of middle-class mothers with the long-term environmental and health consequences of nuclear weapons testing.[21] As a result of their formative meetings in 1957, SANE articulated a position statement on nuclear testing that drew a direct connection between these kinds of maternal concerns over the domestic environment and environmental degradation at large.[22] Their position on the issue was first stated in their 1957 statement of purpose, in which they wrote:

> There is complete agreement on the fact that radioactive strontium falls to earth, it settles like rain. Human beings who drink water or eat food contaminated by radioactive strontium are exposed to its dangers. Children who drink milk from cows grazing on contaminated land are exposed to the dangers of radioactive strontium. There is complete agreement about the fact that radioactive strontium can cause leukemia, cancer, bone diseases.[23]

On 15 November 1957, SANE published the first of dozens of provocative newspaper ads. In a full-page spread in the *New York Times*, they demanded: "No contamination without representation."[24] Their campaigns to raise awareness, however, increasingly targeted maternal concerns—from highlighting the presence of strontium-90 in children's teeth to utilizing the influence of child-advice expert Dr. Benjamin Spock. The 1962 full-page ad depicted a concerned Dr. Spock with a statement that read "I *am* worried. . . . As the tests multiply, so will the damage to children—here and around the world."[25] The Spock ad appeared in over eighty newspapers around the country that year, sending mothers the message that environmental degradation fell under their domain.[26]

Public awareness campaigns against nuclear testing by SANE initially focused exclusively on cow's milk contamination, but it was not long before mothers' questions about breastfeeding safety reached La Leche League. The League's Medical Advisory Board, consisting primarily of pediatricians and obstetricians, agreed throughout this period that breast milk remained a more pure source of nutrition than cow's milk with regard to nuclear contamination. La Leche League's successful guide *The Womanly Art of Breastfeeding* emphasized that not only was breast milk less likely to be contaminated than cow's milk, but in the event of an actual nuclear disaster it would be the safest and most available food source for an infant.[27]

That the mother's body might act much like a purification filter was an idea reiterated in medical journals as well as by environmental activists throughout the atomic age. A Michigan chapter of the group Women Strike for Peace, an outgrowth of SANE that formed in 1961, adopted this exact position. In a 1962 pamphlet instructing mothers on "precautionary measures," they stressed that "natural breast feeding of infants reduces Sr-90 in infants' milk. The mother's body acts as a screening mechanism for the milk."[28] This faith in the maternal body's ability to protect the nursing infant, however, slowly eroded in light of increasing scientific evidence that suggested the mother-infant pair was alarmingly permeable to environmental influences. By the end of 1962, the Women Strike for Peace Committee on Radiation was reporting that while humans "discriminate against Strontium more than cows," mothers had to breastfeed for at least six months and preferably longer in order for their infant to reap these beneficial effects. Because most breastfeeding mothers did not make it past six months, findings like these had the effect of diminishing the arguments about the superiority of breast milk feeding over formula feeding.[29] While concerns over nuclear fallout prompted Americans to start thinking about their bodies and the natural world in new ways, the real challenge to constructions of mother's milk as natural came from pesticides and water and food contamination.

Discovering Toxic Bodies

As early as 1945 scientists had traced the transmission of DDT through the milk of laboratory animals and by 1951, researchers had found DDT in human breast milk.[30] Laboratory analysis by the early 1950s provided evidence not only that the bodies of human beings were permeable to environmental contaminants, but that these chemicals seemed to have a long shelf life once inside the body. As one group of Food and Drug Administration (FDA) researchers led by Edwin Laug put it in 1951, "The storage of DDT in adipose tissue is an extremely sensitive 'biological magnifier' of trace amounts of ingested DDT." Laug and his colleagues went on to explain that based on animal studies, just "0.1ppm DDT in the diet of the rat leads to the storage of about 10 to 15ppm in the fat." As this research progressed, however, Laug and his team found results that indicated breast milk, too, served as a "biological magnifier" for the chemical. The shift in the scientific view of the maternal body from a natural filter to a biological magnifier had broad implications for how people thought about breastfeeding and environmental toxins.[31]

Through the 1960s, studies showing the concentrations of DDT in human bodies continued to appear in leading medical journals and were increasingly publicized in the popular press. Most notably, Rachel Carson explicitly drew the connection between DDT and breast milk in *Silent Spring*, highlighting that it was through breast milk that "small but regular additions to the load of toxic chemicals" in the infant's body accumulated.[32] The 1968 formation of the Environmental Defense Fund (EDF) also played an important role in publicizing the harmful effects of the pesticide. One of the key players in the EDF was Stony Brook biology professor Charles Wurster, who first outed DDT as an ecological hazard as well as a likely and pervasive carcinogen in a paper he published in *Science* in 1965. With Wurster's assistance, EDF would go on to build a health-based legal case against the use of DDT by the end of the decade.[33] The EDF's 1968–69 arguments about the deleterious effects of DDT notably hinged on the testimony provided by Swedish scientists that DDT levels in human breast milk across the globe were twice the maximum intake level that had been set by the World Health Organization. Building on an already existing tradition from the antinuclear movement's organization SANE, which had explicitly targeted maternal concerns over radioactivity in cow's milk, the EDF took immediate advantage of the breast milk contamination issue—hitting newspapers across the country with an add ominously asking, "Is Mother's Milk Safe?"

The public outcry that followed as a result of the EDF campaign placed the breastfeeding advocacy and support group, La Leche League, in the middle of a national crisis. A 1969 League newsletter reflected upon the onslaught of concerned letters and phone calls, concluding that

DDT was "not only poison"; it was also "the number one headache at the LLLI office!"[34] The League staunchly defended the superiority of breastfeeding in the face of this crisis, and in doing so, emphasized that in breastfeeding, women were doing more than simply providing food for their babies—they were embodying natural motherhood, and that, they argued, was worth protecting precisely *because* human bodies everywhere were being assaulted by unnatural chemicals. In articulating the defense against the onslaught of scientific evidence suggesting breast milk was less than pure, and perhaps not quite as "natural" as many had thought, the League and other breastfeeding advocates stressed the ideology of natural motherhood and the holistic benefits of feeding infants at the breast over the chemical content of the milk itself.

The environmental contamination of human milk was the only overtly political issue that the League ever openly supported.[35] While officially La Leche remained unaffiliated with political groups or movements, the fact that it actively encouraged mothers to contact their government representatives is noteworthy and speaks to the degree to which the organization overlapped philosophically with these other "natural" movements. In one of its most coordinated activism efforts, La Leche League groups around the country set up booths, passed out buttons, and marched with signs on 22 April 1970, the first Earth Day, in an effort to raise awareness about the issue of chemicals in breast milk. Concerned that the recurring news about toxins in mother's milk would dissuade some women from breastfeeding, the League took the Earth Day opportunity to remind women everywhere that "breast milk is still baby's best milk."[36]

To the League's dismay, however, the DDT issue only continued to grow more serious as it expanded to include a list of toxic chemicals that ignored the human/nature border. In 1976 the EPA validated the wave of concern over breast milk contamination when it announced its findings of a host of known toxins present in the breast milk of over one thousand women tested from around the country. The report mapped out a new geography of a toxic United States. Using women's bodies as data points, the results constructed a frightening landscape of bodily permeability and contaminated breast milk. Regional variations emerged that paralleled and conformed to the agricultural, industrial, and chemical landscapes in which women lived. Kepone, the active ingredient in the fire ant poison Mirex, marked the bodies and milk of women throughout the southern states; women throughout the Great Lakes region were linked together by inordinately high amounts of PBBs and PCBs; and women in northern California and Oregon found their milk laced with a forestry-related herbicide known as 2,4,5-T.[37] Across the nation, nursing mothers learned their milk was a reservoir of modernity's technological by-products—chemical monsters with names like DDT, dieldrin, oxychlordane, and heptachlor epoxide.

La Leche League continued to find itself in the position of maintaining breastfeeding's good image even while the organization, too, grew more concerned about the situation. A 1976 newsletter captured its continuing resolve in light of another chemical scare. "The presence of these environmental contaminants in mother's milk serves as a warning signal that they are present in all humans," the League wrote.[38] Mothers' worries over environmental pollutants in mother's milk remained such a persistent issue that the League eventually devoted an entire publication to the subject, "Environmental Contaminants in Mother's Milk." "What's a mother to do?" the League wrote, before suggesting women reduce or eliminate their use of pesticides and other household sprays, grow pesticide-free vegetables in their own garden, peel store-bought produce, cut the fat off meat, avoid laundry products that contained "suspect chemicals," and stay away from chemically treated clothing. Finally, they added, mothers needed to write to their "senators, congressmen, and local officials supporting laws aimed at protecting the environment."[39]

Despite League efforts to manage the discourse surrounding toxins and breastfeeding, the growth of the environmental movement helped expose the presence of more and more hidden chemicals in nature and in bodies. The more women learned about these issues, the more they questioned the rhetoric of "the natural," and debates over breastfeeding emerged, anew. "Friends kidded me about being a crank when I was pregnant," wrote Glenda Daniel, a reporter for the *Chicago Tribune* in 1977. But despite the "heroic effort" it took, she successfully "avoided aspirin entirely" throughout her pregnancy and stayed far away from artificial sweeteners and any other unnatural food additives she identified. All these precautions, however, seemed suddenly for naught when Daniel learned five months after her daughter was born that "none of those commercial products I had so carefully avoided has as many noxious chemicals as I carry in my own body."[40] Daniel traced her own concerns to the early 1960s, when she had read about "the pervasiveness of cancer-causing pesticides and industrial chemicals (like DDT and PCB) in Rachel Carson's 'Silent Spring.'" At the time, however, she found the information about bodily contamination horrifying in an abstract and somewhat detached way. In the wake of the 1976 EPA studies and the 1977 U.S. Senate subcommittee hearings, Daniel embraced the emerging new public health–infused mantra of environmental action through bodily containment when she sent her breast milk to a laboratory to have it tested. The results came back and rocked her. Daniel learned that her milk was indeed riddled with a list of chemical components that read like an industrial chemical company's inventory. Simultaneously, the samples she had sent to the same lab from a "can of infant formula" and a "quart of homogenized whole milk" each yielded notably fewer toxins than her own milk. "I suppose I'll wean her now," she wrote of her daughter. "Why take a chance?"

Daniel's report provoked a response from the husband of a La Leche League founder as well as an obstetrician, Dr. Gregory White. White publicly denounced Daniel's decision to "deprive her baby of the certain benefits of breastfeeding to avoid some purely theoretical dangers of environmental pollutants."[41] He also stressed that there had been no "known case of a baby damaged by contaminants in breast milk" compared to infant formulas, which he argued "do contain more of a common contaminant which has damaged children—lead."[42] This, in turn, prompted a rebuttal from Daniel who wrote that upon contacting the League for an official statement on its position on toxins in breast milk, she had been "surprised to learn that La Leche League does not favor informing mothers of the facts about industrial agricultural pollutants." The League spokesperson who had communicated with the *Tribune* claimed that the issue was "too complex for most mothers to comprehend."[43]

While breastfeeding advocates feared that the issue of toxins in breast milk would dissuade women from doing it, breastfeeding rates rose in the 1970s for the first time in decades.[44] The resonance of breastfeeding with cultural imaginings of the natural world as a buffer against the unwelcome encroachment of pollutants persisted in spite of the proliferation of breast milk contaminants, whose presence suggested that, given the onslaught of medical support for infant formula, breastfeeding was not so superior.[45] As the percentage of breastfed babies rose, women struggled with the implications of attempting to live, mother, and breastfeed "naturally" in the modern world. And as mothers continued to seek empowerment and purpose through the pursuit of an identity tied to nature, many women discovered that living a "simpler life" was not always so simple. The burdens that women faced when they pursued a life informed by ideologies of "the natural" seemed unduly heavy, something many breastfeeding mothers, as much as back-to-the-land homesteaders, realized when they found life in the woods or on a farm or with an infant at their breast far less romantic in practice than in theory. Additionally, as the exchange between Daniel, Dr. White, and La Leche League suggests, women who identified themselves as educated and thoughtful mothers who wanted to do "what was natural" by breastfeeding were becoming increasingly alarmed over the threats of toxins in their breast milk. While La Leche League could come across as elitist and antifeminist on this issue, it in fact tried to articulate a message that focused on a bigger picture than the one Daniel explored. The argument between individual choice and the more maternalist rhetoric epitomized by the League did not remain confined to debates about environmental contamination. Instead, the issues that emerged through the toxic breast milk phenomenon merely suggested a much larger divide over the best route to women's empowerment that was forming in the growing back-to-the-breast movement.

Feminism and the Ideology of Natural Motherhood

I personally believe that LLL has always been a part of the women's movement—in the large sense. We don't of course belong to that segment of it which seems to imply that the woman who chooses motherhood as a special role in her life, and wants to devote herself to it, is a poor wretch who needs to be liberated. LLL has helped women to speak out and stand up for their rights as mothers. And because we in LLL support each other, many have gained confidence in themselves and thus have learned to be assertive in an effective and appropriate way. —*Edwina Froehlich*, La Leche League founder, in a letter dated 30 July 1977

In Eleanor Agnew's semiautobiographical account of the back-to-the-land movement, she tells the story of Pam Read Hannah, a self-identified "hippy" who lived on a commune with her husband. In 1966, before arriving on the commune, Pam had "delivered her first child by herself at home on a comfortable floor mattress." The birth went well, but her resolve to breastfeed did not lead to the same happy result. "I went to a doctor finally because I was so miserable," she recalled. She sought relief from the pain of engorged breasts and fever after just two weeks of nursing.[46] Desperate for help, Pam visited a doctor because she felt she had no choice: the experience left her feeling "less than" and her breastfeeding attempt was cut short. Living on the commune with her family by 1968, however, she felt more convinced of her "natural" abilities to birth and nurse her second child. "Alone in the tent [on the commune] with her husband, she knelt on her hands and knees when the moment came, and pushed out her new baby daughter."[47] Again, however, infection interfered, this time in the form of potentially deadly puerperal fever, and Pam had to seek medical attention soon after the birth. Her illness and subsequent treatment thwarted her hopes of breastfeeding. She reported not only feeling like a failure because she had not been able to successfully breastfeed her first son, but also because she was forced to seek help from medical experts in both cases. She had been unable to birth and breastfeed her infants on her own.

As more women became enamored with the opportunities for empowerment offered by the ideology of natural motherhood, many began to question the implicit arguments about the gendered division of labor and risk that seemed to accompany the pursuit of the "simple life." Even as more women sought the experiences of breastfeeding and natural childbirth, the ideology of natural motherhood began to crack under the weight of feminist reconstructions of female roles. As one homesteading, breastfeeding mother put it in 1979, "I am exhausted. I want *my* mother, or canned soup, or baseboard heat. Or do I? At least if I could switch on a light when baby cries at 2am, not fumble with a kerosene lamp while milk leaks down my chest and baby howls . . . I'm so tired."[48]

In the early 1970s, breastfeeding rates in the United States were exceedingly low, with some indicators reporting that only 22 percent of mothers were even initiating breastfeeding in 1972. By 1984, however, the number of women who initiated breastfeeding in the United States had climbed to near 60 percent (an average of state data)—a staggering increase given decades of decline.[49] This generation of back-to-the-breast mothers emerged in the midst of environmentalism and the back-to-the-land movement, but they were also exposed to late twentieth-century feminism, with its conflicting ideas about the experience of motherhood represented by the philosophies of equality and difference. The philosophy of equality feminism in many ways posed a direct challenge to the relationship between breastfeeding and the ideology of natural motherhood by arguing that the construct of gender difference that the ideology relied upon was, in and of itself, a tool of patriarchal oppression. As historian Wendy Kline has recently articulated in the context of a discussion of natural childbirth and feminism, equality feminists of the era "sought to transcend that biological barrier [i.e., the female body] by de-emphasizing the body."[50] Thus, by emphasizing women's physical differences and highlighting them as essential, breastfeeding could be easily co-opted by those who wished to limit female autonomy and independence because it seemed to require that women remain home with young children.

The ideology of natural motherhood stressed women's biological and reproductive functions as unique and essential to a female identity, an emphasis that could have both positive and negative implications. Indeed, many supporters of breastfeeding from the 1940s and 1950s had adopted the ideology of natural motherhood as a way to buttress conservative arguments about "family values" that stressed the perfection of a hetero-nuclear family with a breadwinning father and a stay-at-home mother.[51] The "difference" feminism that emerged in the 1970s, on the other hand, most famously articulated by poet Adrienne Rich, argued that while the "feminist vision has recoiled from female biology," women might stand to gain even more by embracing their biology. She called upon feminists to "view our physicality as a resource" and argued that such physicality could and should be wielded "into both knowledge and power."[52] This perspective sat at the heart of the efforts of the Boston Women's Health Book Collective (BWHBC) and spurred women in the 1970s and 1980s to seek empowerment through the knowledge and experience of their biological abilities. Women's interest in informed and educated childbirth grew in these years, and not surprisingly, interest in breastfeeding increased as well. Mothers, seeking to "touch the unity and resonance of [their] physicality, [their] bond with the natural order," gravitated toward the experiences that the ideology of natural motherhood emphasized, particularly breastfeeding and natural and home births.[53]

For many women who embraced a feminist philosophy in the 1970s and 1980s, breastfeeding appealed to their desire to feel connected to "the experience of being female."[54] As a woman-centric ideology, natural motherhood could encourage a multifaceted model of female sexuality that emphasized the breadth of women's sexual natures, just as the work of scientists like Margaret Mead and Niles Newton had suggested. Natural motherhood was not easily categorized as inherently "feminist" or "not feminist." As a result, women in general had a difficult time deciding where breastfeeding fit into the politics of the women's movement.[55]

As many mothers discovered, the romantic, idealized vision of the "simple" and "natural" life could eventually wear away under the gendered realities of day-to-day life. In addition to the futility of trying to escape the reach of a widely contaminated environment, escaping within a "natural" existence often raised tensions over relationships in the family. The gendered division of labor that seemed implicit within the ideology of natural motherhood grew less and less appealing to this generation, galvanized by feminism and liberated, in some sense, by the lessening stigma of divorce. As one back-to-the-land participant recalled: "You wish *he* would do the laundry for once! Why is it always your duty to stand for two hours hand-pouring water into the machine, stuffing the filthy clothes in, and feeding them through the wringer? But of course, he's too busy with more vital tasks—chain-sawing the wood or fiddling with the dead car."[56] As this quote suggests, the romance of "the natural" faded for many women as a result of these trials and tribulations. The issue of breastfeeding was not quite so straightforward, however. Many mothers found that they liked breastfeeding. Much as La Leche League had always advocated, it could and did provide many women with a sense of enjoyment, purpose, and accomplishment. Many women continued to feel that it was an important part of their identity as mothers. With the help of the counterculture and the widespread embrace of the ideology of "the natural," breastfeeding rates soared in the 1970s. While longtime breastfeeding advocates were pleased, the back-to-the-breast community faced a more diverse and educated population of mothers who demanded room for personal choice and individual circumstances when it came to recommendations to breastfeed.

With more educational resources at their disposal than previous generations, mothers in the 1970s and 1980s struggled with what breastfeeding meant for them, personally. Was breastfeeding feminist? The answer was ambiguous at best, and while the members of the Boston Women's Health Book Collective overwhelmingly chose breastfeeding for themselves, they were not proselytizers on the subject. "We did it because we wanted that experience, and also because we were feeling proud of our bodies and glad as women that our bodies can provide nourishment for our children," they wrote.[57] Unwilling to dictate to their readers, the authors of *Our Bodies, Ourselves* followed their brief breastfeeding endorsement with a laundry list of reasons why a woman might very reasonably choose *not* to breastfeed, including: "We are told that our breasts are our sexiest parts, and . . . many of us feel embarrassed or uncomfortable using our breasts to feed our babies," they explained.[58] Additionally, they were quick to point out that "sharing child care is easier with bottle-feeding," adding that "it's difficult to establish a successful part-time breast-feeding arrangement, but if you want to have time for yourself away from the baby, it's worth the effort."[59] Most importantly, they suggested that bottle-feeding would diffuse the "exclusive power to meet your baby's needs" for those women who found the situation of "total mothering" to be too overwhelming or unfulfilling.[60]

Such an apparently neutral approach to the decision to breastfeed, of course, was in conflict with the philosophy of La Leche League. In the 1979 edition of *Our Bodies, Ourselves*, the BWHBC explicitly discussed its perspective on the League and its book, *The Womanly Art of Breastfeeding*, which remained a formidable force in the breastfeeding community even as other successful advice books became available.[61] "*The Womanly Art of Breastfeeding*," the Collective wrote, "will give you facts and confidence . . . [and] it will answer any specific questions you might have." "But," they cautioned, "its philosophy is different from ours. We do not believe that breast-feeding has to dominate your life."[62]

Founded in 1969, the Boston Women's Health Book Collective was similar to La Leche League in that it was a female-run, woman-centered group dedicated to empowering women through knowledge about their bodies. The women who founded both groups did so in order to gain knowledge and insight into subject matter that had been professionalized out of their reach by organized medicine. Much like women's nursing support and childbirth education groups, BWHBC began as a series of meetings in which the members researched specific topics and then shared their findings with the group. BWHBC, however, differed in important ways. Unlike the women's education groups of the previous decades, its goal was the inclusion of many different perspectives, and the Collective focused on a host of issues that women faced, not just those pertaining to motherhood. By the end of its first series, BWHBC made its resulting course available to women's groups around the country. The end product became, of course, the now internationally known book *Our Bodies, Ourselves*.[63] BWHBC helped to spark the practice of "consciousness raising" for women through bodily awareness, education, and self-discovery. The Collective and the women's health movement that it helped fuel took various forms, but it is now looked upon as a constitutive component of the politics and activism of the second-wave feminism that characterized the 1970s and 1980s.[64] The women's health movement quickly inserted itself into political discussions on everything from birth control safety to abortion rights.

Consciousness raising, the idea that learning about one's own body could have political implications, sat at the core of the women's health movement. This orientation, however, was not something that the earlier breastfeeding reformers had ever explicitly adopted or articulated. La Leche League, for example, believed that individual empowerment could occur through the *natural experience* of a woman's ability to feed her children, which would lead to personal fulfillment, which would lead to happier families ruled not by medical authority, but by the authority granted to women by nature. As League founder Mary White put it, "Doing what comes naturally turns out to be the best thing we could be doing because it really helps us to bring out the best in our children by bringing out the best in ourselves."[65]

BWHBC's approach to female empowerment, however, emphasized instead that individual autonomy was rooted in the process of obtaining and wielding *expert knowledge* about one's own body. The Collective believed that it was key for women to hold authority over the construction and implementation of scientific and medical knowledge about their bodies so they could break out of the patriarchal patterns of American society and attain self-fulfillment. "We learned what the 'experts' had to tell us," they reflected, only to find that the "rote memorization" and "abstractions" they encountered in reading traditional textbooks and medical articles continued to produce a distance between their "selves" and their "bodies." The knowledge about their bodies and their selves did not "really become our own," they wrote, "until we began to pull up from inside ourselves and share what we had never before expressed." The process of making this connection was central to becoming empowered, freed, liberated, and growing into "better *people*."[66] In many ways, as historian Lynn Weiner has observed, La Leche League's position predicted the interest of the women's liberation movement in health care when it articulated a female role that usurped medical authority over the female body, but the League never asserted this to the extent that the women of the BWHBC did.[67]

The ideologies of natural motherhood and second-wave feminism clashed over what it meant for a woman in American society to reach her "full potential." The ideology of natural motherhood, of course, suggested first and foremost that a maternal identity was a necessary component of mature womanhood. For many mothers on an individual level, the draw of breastfeeding stemmed from the resonance of the idea that the female body could connect to a timeless, evolutionary identity through its biological functions. The search for a fully realized maternal identity encouraged the experience of motherhood in all its biological complexity, including breastfeeding. "Breastfeeding," one woman wrote, "does nothing for women's liberation but enhance it!"[68] These feminist-influenced mothers who sought breastfeeding and a connection with their "natural" selves, however, often realized that they, too, had conflicting feelings about some of the implications of tying breastfeeding to the ideology of natural motherhood. Adrienne Rich's 1972 critique of the "natural mother" concept captured the frustration that many mothers struggled with when confronting the realities of breastfeeding. It was an "unexamined" assumption, she wrote, "that a 'natural' mother is a person without further identity, one who can find her chief gratification in being all day with small children, living at a pace tuned to theirs; that the isolation of mothers and children together in the home must be taken for granted; that maternal love is, and should be, quite literally selfless."[69] Rich herself, like the BWHBC women, breastfed her children, but the intense, physical relationship required deep introspection in order for her to make sense of the complicated dependence she experienced with her infants.

Mothers during this era struggled against the backdrop of the feminist movement to come to terms with the tensions that breastfeeding revealed. Was breastfeeding an outmoded and potentially antifeminist way of doing things? Or was it a progressive, self-affirming, feminist choice that subverted a male-centric view of the female body? Did it, as Rich suggested, provide women with an opportunity to experience a "new realization . . . of her being-unto-herself"?[70] Whether they were allied with La Leche League, BWHBC, both, or neither, mothers who sought to breastfeed during the 1970s and 1980s faced a different ideological landscape than had earlier generations. Just as American women everywhere struggled to make sense of their evolving political and personal identities, mothers who breastfed sought to define themselves in light of the era's challenge to domesticity and expert authority.

In a 1981 episode of the highly popular daytime television program

The Phil Donahue Show, the tensions between natural motherhood and feminism were brought to light when Mary Ann Kerwin, a founding member of La Leche League, appeared alongside Norma Swenson, founding member of the Boston Women's Health Book Collective. In front of a live studio audience filled with nursing moms and their infants and toddlers, Donahue lightheartedly worked the room. "We clearly have a number of people in our audience who are much more impressed with Big Bird than they are with me," he joked. The format of *The Phil Donahue Show* encouraged audience participation, and the audience consisted of two hundred members of La Leche League, all of whom were in Chicago that week for the League's twenty-fifth anniversary convention. The guests and the composition of the audience made for a lively discussion. One mother commented that her involvement with the League and its support of her breastfeeding helped her "to become a better mother," by teaching her that "children are people and they have real feelings." Donahue pressed her on the connection she drew between breastfeeding and maternal empathy, asking, "Can't you do that with Enfamil, too, though?"—a reference to manufactured infant formula. "Not to the extent," she replied. "[Breastfeeding] brings you a lot closer to your child and, therefore, you get a lot more patience to deal with them."[71]

Amidst this crowd of La Leche supporters, BWHBC founder Norma Swenson proceeded with caution when Donahue asked her to address her concerns with La Leche's approach to breastfeeding. She began, "I think La Leche is perhaps the most successful example of a women's self help group in this country," Swenson lauded. "It certainly must be the first one and it certainly has shown that women who have a common experience together, which share information and support can build a better mousetrap—can actually make it more satisfying for women than the advice that they can get from experts and that's what self help is all about." When pressed further, however, Swenson delved into the crux of the conflict between feminist ideology and the League's approach to breastfeeding. She replied:

> We possess this kind of image and sometimes I think the League does this without really intending to, that a woman really has to be a mother and stay home or go out to work. . . . The truth is that one woman in ten is the head of a household in this country and that over half the women that work are

> mothers. . . . How are we going to make it possible for women in the work place to breastfeed and particularly poor women. . . . I want La Leche to be a feminist organization.[72]

Swenson's assessment of the League and its position on working mothers was generous. In fact, the League fully intended to convey the message that being home full-time was always better than any alternative. Swenson's comments, however, also highlight the feminist position that women should have the choice to breastfeed or not to breastfeed and that decisions about reproduction and child rearing should not be met with moral judgment. Working mothers were prime subjects for a debate of this nature because so many women who worked did so because their economic position demanded it. Swenson's remarks suggested that whether for financial reasons, marital status, or mental health, women worked because they needed to. By forcing hardworking mothers to feel guilty because they could not breastfeed, Swenson argued, the League was disempowering an already marginalized group. For breastfeeding advocates, however, the overtly political rhetoric of choice had no place in the breastfeeding debate.

Mary Ann Kerwin responded, "Norma, first of all, I don't think helping breast feeding mothers who are employed outside the home is a political issue and we are helping them. . . . We have an information sheet on how to help mothers who are employed and working." She added further that "just now the Board of Directors reevaluated our purpose which is to help all mothers breast feed and to implement that status we are initiating pilot projects with employed mothers . . . so we are trying to do more of it because, of course, we agree with you on the statistics."[73] The League did reach out to working mothers, but for years they did so with the underlying goal of convincing women that staying home was far more fulfilling than trying to pursue a career and motherhood simultaneously.[74]

Donahue asked, "Is it your suggestion that the nursing mother has, however wonderful and beautiful, lost just a little more freedom in this pursuit of self after the birthing?" To which Swenson replied, "I think that the two don't have to be incompatible. I think that a woman can have a sense of herself but she needs to take care of herself. She needs to watch out for burnout and to let other members of the family help out. You know, I don't like to say this in a place like this, but there are advantages to bottle-feeding and one of them is that other members of the family can take care of the baby in the most satisfying and intimate way by feeding it." At this point, a La Leche League member of the audience stood up and commented, "You can fill up a bottle with your mother's milk . . . and then somebody else can feed them the mother's milk. So that really isn't a problem." Other League audience members had much more to say in response to Swenson's comments. One mother said, "You say that the family can also get involved through the bottle-feeding. Well, my husband has been able to get involved intimately with our child by bathing him and playing with him and keeping him busy while I'm cooking the meals, so it's not just the bottle [that] can do that."[75]

In response to Swenson's statement about a mother's sense of self, one mother replied, "I never got closer to myself or found out more about myself and felt more fulfilled than when I began breastfeeding. Through La Leche League I had so much information and learned so many things that it was the most beautiful thing in the whole wide world to me and I loved to share it." At this, Mary Ann Kerwin interjected, "Phil, that was what I was trying to say. Twenty-five years ago we tried to give women the right to do what they wanted to do. They wanted to breastfeed and they were put down and told they couldn't do it."

Donahue redirected the conversation back to gender roles specifically, saying that "La Leche, I think, is also perceived by some feminists as an organization with—however wonderful and committed to its goals many of them achieve, also has implied in it the notion that somehow the woman is the nurturer, the homemaker, the cake baker while in the traditional role of lots of mother and not much father." To which Kerwin replied, "No," and Swenson stated: "There is a failure on the part of the League to take an active role in making sure that women have a choice about whether or not to give birth at all. . . . I think the League should be in favor of choice for women about abortion rights. I think the League should be in favor of the Equal Rights Amendment. I think the League should be involved in encouraging the teaching of the birth control and sex education [*sic*]."[76] To this, Kerwin replied, "Norma, we just don't have time for those issues. We are so busy with the breastfeeding mothers who need our help."[77]

The dialogue between Swenson, Kerwin, and the audience in this brief interchange represents many of the central issues that came to characterize the nature of the breastfeeding debates between feminists and strict adherents of natural motherhood ideology as La Leche League articulated it. The ideological gaps between the feminist women's health movement, embodied by the Boston Women's Health Book Collective and the ideals of La Leche League, in particular, are highlighted in these transcripts. Letter responses to Swenson's appearance on the program expressed broad consensus among women in the women's health community that while the League was doing an excellent job supporting women who wanted to breastfeed, it was "very conservative otherwise."[78] Disagreements over the position of the League on political activism, women's rights issues such as the ERA and abortion, and even on bottle-feeding and working motherhood rippled throughout the breastfeeding community.

La Leche League members and leaders alike devoted much time and energy, as a group and individually, to working toward reconciling the new feminism with the ideology of natural motherhood that had helped fuel breastfeeding's return. For many women like Patty, whose story introduced this chapter, the broad ideological differences of the opposing sides of the debate were largely meaningless when it came to actually living one's life: "I don't think I felt a part of Women's Lib. My feeling was you're an individual, pick what you want, pick and choose. . . . I think that for the most part, I carved my own way."[79]

Deciding whether or not one was more or less allied with ideologies of modern feminism or natural motherhood often became a nonissue for mothers, whether they breastfed or not. Often, doing what worked when it came to infant feeding defined a woman's approach to motherhood more than anything else. Women's letters to La Leche League and to BWHBC on the subject more often than not had more to do with needing basic help and advice than with wrestling with concepts, theories, or political arguments. As one mother-to-be explained in a letter to the BWHBC, she planned on nursing her baby "full-time," but because of "financial reasons," she would be returning to work after just two weeks of maternity leave. She closed her letter asking for "any help you can possibly give," so that she could make sure her baby had "only breast milk."[80] In many ways, local-run groups like Nursing Mothers grew up to fill the practical gap between feminist and natural motherhood ideology. Focusing primarily on the health benefits and with a less ideological approach, Nursing Mothers groups, often affiliated with local childbirth education organizations, offered women lactation support and knowledge without demanding that women uphold particular ideals of womanhood. The position of these middle ground women formed in relationship to the era's polarizing debates over breastfeeding, motherhood, and women's rights. In the words of Tammy, one of the first hospital-based lactation consultants and a leader of Patty's Nursing Mothers group in the 1970s, "In my day, in the . . . '70s, La Leche meant you breastfed forever, you never left the house, everybody slept together in the same bed and god forbid you would go back to work, or want to go out without a baby on your breast and so it developed the name of 'these are fanatics.' Nursing Mothers didn't carry that. It was 'we'll support you whatever you want to do.'"[81] Tammy's assessment of La Leche League became more and more widespread, and accurate, in this period as the League's construction of natural motherhood through breastfeeding became unattainable for, and unwanted by, a growing majority of the women who chose to breastfeed. The ideological gaps that could exist between feminism and natural motherhood did impact women's lives when it came to breastfeeding. Nowhere was this effect more visible than in the struggle hinted at by Tammy, over breastfeeding and working mothers.

Being "Better" Mothers

In 1978, a mother from Missouri, who also happened to hold a PhD, wrote to the League's board of directors. Offended by La Leche League's "alienating" philosophical positions on motherhood, she wrote, "La Leche League certainly has a right to its own philosophy of 'mothering' (I would say 'parenting', but you rarely mention fathers), but it is my opinion that the organization's *primary* service is to provide technical, 'how-to' information which is surely needed."[82] Critiques like this grew throughout the 1970s, indicating an increasing amount of discord within the breastfeeding community. In what would become a characteristic response, the League's executive secretary, Edwina Froehlich, replied firmly that the mother's "angry reaction" was most likely "due to a misunderstanding" of the League's "primary purpose." "We are dedicated to good mothering through breastfeeding, not just to providing technical how-to information about breastfeeding, as you thought," she wrote, adding, "We believe that understanding the *mothering* an infant needs is most important."[83]

In separate, internal correspondence among League board members regarding this letter, one member emphasized that ultimately "LLL cannot be all things to all people," adding that it was clearly defined in the organization's philosophy that it encouraged "total mothering," a mothering approach that was in distinct opposition to the mothering offered by "cultural mothering," a model in which a woman believed it was "normal" for her "to be away from her children for varying times and reasons."[84] While the letter's author had ended her correspondence with a gesture of sisterhood—"We're all on the same side—*all* of us want the very best for our children and for ourselves"—the LLL board saw little common ground with her, or women like her. "We stress that the role of the mother is to see that her baby's needs are filled. If his need is for her presence, he will certainly suffer from her absence." In private, the board discussed that in fact LLL's position was unapologetically "to discourage mothers of the very young from working away from their little ones."[85]

In spite of (or perhaps because of) their carefully thought-out responses to letters like this, the League became quite adept at infuriating feminist-leaning women over these years. Its stance on the issue of mother-infant separation could at best appear cool and unfeeling toward working mothers and at worst could seem outright judgmental and offensive. In retrospect (as well as to many women at the time), the League's somewhat antagonistic attitude toward working mothers can appear to be in marked contrast to its original approach of offering a supportive and reassuring place for women to learn how to breastfeed effectively. What had long been an internal organizational dialogue over breastfeeding and mother-infant separation became, in the face of second-wave feminism, an increasingly polarizing debate linking breastfeeding to the politics of working motherhood. In a circa 1974 draft of their leader-policy manual titled "Outreach," for example, the League included a section called "The 'Liberated'(?) Woman . . ." which advised leaders on how to respond to "an attack from a woman who feels very strongly that children should not tie one down."[86] They provided five potential responses that leaders could utilize in addressing the mother who wants "a career outside the home," but in each scenario League leaders were to remain firm on the point that the ideal situation was for mothers to stay home with their children until they went off to school.

While the League had always maintained this fundamental belief at the core of its construction of natural motherhood, the degree to which its leaders had been challenged to defend it increased over the years. In 1969, for example, a League leader from New York wrote in the newsletter *Leaven* on the subject of League philosophy and membership: "Another subject which has come up is just how far we can 'push' LLL philosophy." "Well," she replied, "we don't push it at all—except," she added, "when it comes to prospective leaders." She later emphasized it was important to remain nonjudgmental when dealing with potential League members or mothers who were simply looking for some advice. "We are never 'nosey' here," she stressed. "What [a mother] decides to do, whether regarding bottles, solids, weaning or childbirth, is none of our business."[87]

This low-profile approach to mentoring women had worked quite well for the League for years. The League's founders believed wholeheartedly that most women who experienced successful breastfeeding would come in time to realize their "natural" roles as mothers. A thriving breastfeeding mother-infant pair, they reasoned, required the presence of the mother at almost all times anyway. Thus, League leaders hoped that the "problems" of working motherhood and mother-infant separation would address themselves once more women breastfed their infants. As one founder put it, breastfeeding was simply a "means to an end."[88] Over time, however, as the League grew, the LLL board found it increasingly difficult to maintain this core philosophy with such a relaxed hand. In reality, the League headquarters could know very little of what went on in individual group meetings all over the country, and the world, and the board became increasingly uncomfortable with the ways that individual leaders interpreted the organization's policies. As critiques from feminists mounted, as demands for help from mothers increased, and as LLL women became identified as breastfeeding "fanatics," the League struggled to maintain its public image as a benevolent organization of female empowerment while also adhering to its fundamental beliefs regarding natural motherhood and breastfeeding.

In 1980, one woman wrote to the League to critique *The Womanly Art of Breastfeeding* in detail. In a scathing and well-articulated letter, the author took issue most with the book's foreword, written by one of the League's earliest and most avid supporters, Dr. Herbert Ratner:

> In Dr. Ratner's Forward [*sic*], he praises the La Leche League for being supportive "in an age when indices to a sick society are high." Among his sickies are illegitimacy and divorce. You tell me: is it sick to choose (with all the current methods of birth control available) to have and raise a child while not being legally bound to the other parent? And is it sick to divorce a man who is unkind to you and/or your children, or is insensitive to the family's needs? In what way is a divorced mother in a less desirable position than a mother who is trapped in a stifling marriage?[89]

Most offensive to this particular letter writer, however, was the statement "A woman's place is in the home." She wrote, "How dare you make that assumption?! I am appalled!" Disgusted with the League's oversight in allowing such an outdated preface to remain in the 1978 edition of the book, the letter's author closed by writing:

> At this time, your book is insulting to all women in any society. We want to be loving mothers in our own special ways without being expected to fit into an unrealistic degrading role of cheery housewife/mother. . . . Please leave your attitudes about the ideal mother where they belong—out of print.[90]

What non-League women, like this mother, perhaps misunderstood when they attempted to "set the League straight" on the advances of the women's movement and feminism was that the League had not *accidentally* overlooked the continued presence of Dr. Ratner's contribution; it's leaders had actively chosen to leave it there. A response from the League's executive secretary, Edwina Froehlich, in 1977 to a similar letter emphasizes this point: "We believe that every baby's rightful heritage is the assurance that his needs will be filled by his parents. Who else should the baby depend upon? . . . The *ideal* arrangement is for a baby not to have to be separated from mother during most of his waking hours—emotionally he needs her around most of the time until he gets pretty close to school age."[91] The League's official position on the issue of mother-infant separation remained steadfast and immovable well into the 1980s.

Despite inspiring guilt and anger in the working mother, the League's philosophy on mother-infant separation had little impact on the group's utility as a store of information and resources dedicated to breastfeeding. Women, pediatricians, and nurses picked up the League's *The Womanly Art of Breastfeeding* regardless of their views on family structure because it offered such practical and proven advice on the subject of breastfeeding. The League's policy on working motherhood, however, did directly impact the women who volunteered as local group leaders. Each local group was organized and run by an officially designated LLL leader. To become a League leader, one had to subscribe entirely and without reservation to La Leche's core philosophies, including the Third Concept, which stated:

> The baby has a basic need for his mother's love and presence which is as intense as his need for food. This need remains even though his mother may be absent for a period of time for needs or reasons of her own.[92]

Pledging to uphold the League's ideals took on an increasingly rigorous nature as the years wore on and the number of groups increased.

In the face of mounting external pressures, the organization's leadership turned its focus onto its own flock. In a 1969 issue of the group's leadership publication, *Leaven*, the New York chairman, Barbara Kennedy, defended the League's position on the Third Concept, arguing that the League did not "push" LLL philosophy on mothers—with the exception of group leaders. For League leaders the standards of maternal conduct were necessarily quite different, she argued. Kennedy's defense stressed that the League needed to "make sure that those who represent the League really believe in its ideals." The personal lives and choices of the leadership had to adequately represent the League's goals. "It does matter if a prospective leader started her baby on solids at three months (after all, she has missed some aspects of the nursing experience)," she explained. "But," Kennedy added, "the more important thing is whether she becomes convinced . . . that waiting until at least four to six months is really better." This was all very important to good leadership, Kennedy stressed, because "if she does not, herself, feel confident in such a conviction she will not be able to communicate it to others."[93]

As Kennedy's piece helps demonstrate, the League maintained high standards for its leaders. At the same time, its Catholic roots showed in its acceptance of human failings. Despite high expectations for its leadership, the League's publications emphasized the importance of striving toward an ideal, even if the majority of its membership would never be able to fully embody its collective vision.[94] "Of course we all fall short," one leader acknowledged, "but we also, hopefully, keep learning and maturing." The most important thing for leaders in the early years of the League, therefore, was an acknowledgment of their shortcomings and a willingness to utilize their own experiences—failures or otherwise—in order to help others avoid their "mistakes." Weaning an infant before it was ready, starting solid foods too early, relying on an occasional "relief" bottle, getting a babysitter for Friday night—these were, according to the League, all mistakes, and mistakes were understandable but they were also always regrettable. The yoke of guilt and self-scrutiny that the League used to exert uniformity across its diverse and widespread membership, however, broke down during the 1970s over the Third Concept as League mothers themselves began to challenge this construction of breastfeeding.

In 1977, a longtime member from Minnesota wrote to the League to alert the board to an oncoming storm:

> [The Third Concept] is very mildly worded, I believe, but the meaning is unmistakable. There is no doubt it is the most difficult of all the concepts for many mothers to accept. It flies in the face of our culture's Siren call to "get a job and be fulfilled," "day-care is great," "the quality of mothering is important, not the quantity," etc. . . . Please do not give in to the pressure. [The Third Concept] provides the very crux of LLL philosophy. Without, our manual could just be called *Breastfeeding Mechanix Illustrated*!![95]

She was not alone in her worries. Similar letters from concerned League leaders and members began to arrive into the League's main office in Franklin Park, Illinois as the 1970s waned. Many League followers were even more disturbed by the extent to which local group leaders appeared to disregard the League's policy in their own lives. In 1978, for example, a concerned LLL leader from Indiana wrote to protest the encroachment of working mothers upon the League's leadership. "Frankly," she stated, "I don't see the place of Leader Applicants who are working!" "Life is full of choices," she added. "Moms who wish to represent LLL have . . . made a choice . . . [and] LLL is synonymous with keeping moms and babies together."[96]

The League's board remained committed to the Third Concept, even as critics and members alike railed against it. In a typical response, Executive Secretary (and League founder) Edwina Froehlich replied to the leader in Indiana on behalf of the board, emphasizing the board's continued agreement on the matter: "We agree with you that LLL can choose not to accept Leader Applicants who choose to work and with NO guilt feelings. Such an applicant does not share the philosophy we try to express in our Concept Three, which deals with mother-baby separation."[97] It is difficult to parse the degree to which active members openly challenged the League's philosophy, but letters like those excerpted here suggest that there were significant numbers of mothers connected to the League who felt that their commitment to breastfeeding and natural motherhood ideology was under attack both from without and from within. In response, the organization's leadership held a formal reevaluation of its policies and philosophies. In 1979, after years of informal debate and a full year of intense, organized discussion among League board members, leaders, and members from around the country, the organization reaffirmed and clarified its position on mother-infant separation. After battling with the issue of working motherhood and breastfeeding for nearly a decade, the board announced its "New Third Concept" in *Leaven*, the LLL Leader newsletter. The 1979 version of the principle stated:

> In the early years the baby has an intense need to be with his mother, which is as basic as his need for food. . . . More clearly than before, the Board believes that this new wording puts the League's emphasis where it was always meant to be. . . . Unequivocally stated—we reaffirm this enduring belief.[98]

Simple, yet incendiary in its implications, the 1979 reaffirmation of the Third Concept helped draw a thick battle line between the distinct definitions of motherhood and breastfeeding that were emerging.

Despite the organization's clear stance on this issue, however, there were League leaders, and many, many more members who chose to ignore the Third Concept. Provoked by the discussions on working motherhood printed in *Leaven* and the *LLL News* in the year leading up to and following the 1979 decision, many of these working mothers and their supporters began to "come out" to the organization in defiance, hoping to initiate a change in the policy. One LLL husband wrote in 1979 to defend his wife's choice to work and serve as a League leader:

> During my wife's early League participation as a leader, we had our own business, located physically away from our residence. By choice, [my wife] would work at the business while I stayed home with [our son], then about 16 months old. . . . It permitted me the time I needed to get to know [our son]; and it permitted [our son] to get to know both his parents on a relatively equal basis. . . . I feel strongly that the League is not facing reality when it demands that its leaders never be separated from offspring.[99]

While the League had long expressed an interest in the father's role in child rearing, it had also always articulated a model of the family that stressed that fathers should serve as supports to mother and baby. This sentiment was clearly summed up in one of the League's other longstanding concepts—the "Father/Husband Concept, " which stated: "The father's role in the breastfeeding relationship is one of provider, protector, helpmate, and companion to the mother, by thus supporting her he enables her to mother the baby more completely."[100] La Leche League, therefore, did not accept the implied argument of the above letter, which assumed that all an infant really needed was a parent. For the League, however, there was no substitute for the mother. The organization clarified its position in its response, stating:

> We do realize that as the baby grows so does his ability to be separated from his mother . . . [and] some babies at 16 months are ready to be left with Dad for short periods. . . . [Our focus is on] the importance of permitting the baby to develop a close relationship with first the mother—then as this relationship takes shape and firms up he will begin to deepen his relationship with dad.[101]

The rigid gender-role definition that the League maintained worked both ways; it restricted women to their domestic roles, but it maintained mothers' authority and control over infant rearing as well.

In 1980 a mother from New Jersey, upset that the League's Third Concept would prevent her from becoming a leader wrote, "I am a working breastfeeding mother who will not become a League Leader because Concept 3 has the nerve to imply that I may lack the mothering skills or may even be a bad mommy because I work." She went on to describe the economic circumstances she struggled to overcome. Married, with one child, the couple's son had arrived when her husband was still in school as she worked full-time to support them. "I worked full-time until the day of Daniel's birth and had promised my employer that I would return to work when Daniel was three months old," she explained. Unable to face the emotional heartache of leaving her son once her three months were up, however, the mother sought evening work: "In this way I could care for Daniel during the day, and in the evening Daniel would be cared for by his daddy." So began this woman's struggle to maintain breastfeeding in a two-working-parent household. "I would nurse Daniel about 4:15[pm] then leave for work at 4:45. At 8:00pm I ran home on my half hour lunch break to nurse Daniel and put him to bed. At 8:25 I was back at work. My workday ended at 1:00am and by 2:30am Daniel was usually up for another feeding. Breastfeeding was not easy for me to continue under these tiring conditions, but I did!" Eventually, she was able to find a slightly better job with a slightly better schedule and, as Daniel got older, he needed less nursing. "Daniel, now two," she reported, "still nurses once a day and I still work at my keypunching job. I have a happy, energetic, secure child who has a wonderful relationship with both mommy and daddy, and I'm happy that things have worked out so well for all of us."

Despite her feelings of accomplishment and success in making everything work out for herself and her family, however, this mother found the League's attitude toward working motherhood distressing and personally offensive. "The League insultingly refers to working moms as part-time mothers, as if we somehow lead two different lives," she reported, adding that working, for her, was a "sacrifice . . . not a pleasure." She closed by stating pointedly that she could not "support a group that implies I may not be a good mother," adding that until the League dropped or amended the Third Concept she would not be attending any more League meetings, nor would she renew her membership.[102]

Aside from the personal nature of these disagreements, the strife that the Third Concept caused within the League highlights the shift that was taking place in the relationship between breastfeeding and motherhood more generally by the late 1970s. Unlike breastfeeding within the framework of natural motherhood, as "a means to an end," as La Leche League advocated, mothers in the 1970s increasingly came to see breastfeeding as an end in and of itself. Divorced from the ideology of natural motherhood, or "total mothering," as the League often referred to it, and the belief that a woman must give herself up entirely to her maternal instincts to be with, feed, and protect her offspring, the meaning of breastfeeding began to shift. Extracting mother's milk using hand-powered breast pumps became part of the working mother's breastfeeding experience, adding to the accessibility of the practice but contributing further to the decoupling of natural motherhood ideology from breastfeeding. As greater numbers of women sought to breastfeed, more and more mothers constructed identities, lifestyles, and families that melded aspects of natural motherhood with aspects of a modern, feminist, technologically sophisticated reality, a reality that was inescapable for most families.

The Choice

In 1976, Dolly, a single, white mother, gave birth to a son "through welfare channels." Shortly after leaving her husband, Dolly had discovered she was pregnant. With no money and no health insurance, Dolly found herself with little choice but to go into the welfare system. After undergoing a cesarean section, Dolly did not see her son for over twelve hours after the operation, despite her repeated wishes to see him and breastfeed him. She recounted:

> I didn't have any advocates there for me, so finally, after a lot of battling, I got the baby and I even got them to move me to . . . where the rich women [stay] . . . a place where the baby would be with me. And I just basically sat in the hallway and made a lot of noise until they brought the baby and they did that, so that was the first time I nursed him, it was way too long.[103]

Dolly's experience as a single, low-income mother contrasts markedly with the rosy portrayal of natural motherhood that La Leche League espoused. Dolly worked full-time in order to survive and to provide for her son. Despite her economic position, however, she felt committed to breastfeeding: "Everything that I had read said that babies, well first of all, it is just an instinctual thing to me, it seemed like the best course of action, but also I guess I supported it with other factors such as this is a good time for bonding, for nurturing, it's healthy for me, as it would be for the baby."[104]

Dolly educated herself on the latest in childbirth and infant care advice by reading books she checked out from the library. Prior to her son's birth, she was also able to enroll in Lamaze childbirth classes, and even though she had a rough time in the hospital, she later encountered a few supportive individuals who encouraged her in her desire to breastfeed. "There were a couple of nurses who gave me some very valuable pointers," she remembered. "I would describe them as old hippie types. . . . They encouraged me to not give up and to just keep going."[105] Once she got going with breastfeeding, Dolly says she cannot remember looking back. While there were difficulties along the way, she took the message of the advice books to heart—"relax and don't give up"—and so she persevered.

Dolly worked full-time while she nursed her first son. Her position in her family's business allowed her more flexibility, perhaps, than many working mothers would have had at the time, and she was allowed to bring her son to work with her and nurse him there. Once he became more mobile, his presence in the office became more of a disturbance so she put him into a day care situation. Living in a city allowed her to find a day care that was close enough to her workplace so that she could walk there several times a day on breaks to breastfeed him. Later that year, she took a different position in the same business, this time as a driver. No longer confined by the rules of an office, Dolly found that she could keep her son with her almost constantly in her new position. "I drove all over the country and unless it was all the way from Pittsburgh to California, I took him with me so then I still nursed," she recalled. Eventually, however, the inconvenience of having to stop driving in order to breastfeed drove her to switch to bottle-feeding. After just a few months of this, she switched him off formula to a diet of solid food that she ground up herself, with a hand grinder.

The similarities between Dolly and Patty, discussed earlier in this chapter, are actually quite striking despite their different backgrounds and economic situations. Dolly was (and remains) an avid vegetarian, a lifestyle she chose due to her concern over the treatment and processing of animals. Additionally, she has always stayed away from processed foods. "I'm basically distrusting of big business and of commercially produced foods," she recalled, adding that her childhood was spent "in a place north of Pittsburgh where there were still farms so . . . I could buy a lot of things from farms and make it myself." Obviously drawn to some aspects of natural motherhood, Dolly strove to breastfeed even as she was simultaneously challenging the definition of "nature" held by increasingly conservative reformers who maintained that working motherhood and natural motherhood were two different things.

As a working mother with economic hardships to confront, Dolly integrated the widespread beliefs in nature and maternal instinct of the era with the reality and constraints of her life. Breastfeeding offered a way to express her belief in the power and healthfulness of "the natural" at the same time that she was deconstructing entrenched notions of "natural motherhood" itself. Working mothers like Dolly were perhaps the most effective in changing perceptions and norms surrounding breastfeeding and motherhood. While middle-class mothers argued about the implications of breastfeeding on a theoretical level, working and single mothers forged new identities of motherhood out of nature and necessity.

The League's inflexibility when it came to the issue of mother-infant separation and working motherhood took its toll on the organization. Unwilling to make room for individualized constructions of motherhood, women not only stopped becoming members; many leaders and members left the organization altogether. Between 1980 and 1984, the number of active League leaders in the international organization dropped from 12,458 to 9,602. Between 1983 and 1988, the number of LLL groups dropped from 4,054 to 2,857. But as this section has argued, it was not simply the Third Concept that alienated mothers and potential leaders. The League's interpretation of the ideology of natural motherhood and its definition of "successful" breastfeeding became increasingly anachronistic as new generations of feminist-minded women and working mothers demanded, either in theory or practice, an expansion of women's roles in society. Rigid interpretations of the ideology of natural motherhood through breastfeeding did not offer women the framework of choices that feminists had come to demand. For most feminists the idea of using guilt to coerce mothers into breastfeeding without exception was simply not fair, nor realistic.

Despite mothers' desires to breastfeed, however, the reality of being both a mother and a worker made raising an infant an increasingly complicated task that involved a network of formalized support, whether in the form of friends and family or in paid child care. Some women worked out arrangements in which they could take their infants to work with them, or they were able to secure affordable and reliable child care near enough to their jobs that, like Dolly, they could leave and breastfeed during coffee and lunch breaks. Even these "lucky" working mothers, however, no doubt faced strain and hardship for their pursuit of breastfeeding against all odds. By the 1980s, however, the emergence of an increasingly available and user-friendly consumer device would upend the debate surrounding breastfeeding, motherhood, and women's place in society: the breast pump.

CHAPTER SIX

Woman's Right, Mother's Milk

The Nature and Technology of Breast Milk Feeding

In 1961 in Libertyville, Illinois, Marie Lawe gave birth to a healthy baby girl, Diane, who weighed just under seven and a half pounds. Despite the reassurances of friends and relatives that her baby would "cry when hungry," Marie knew almost immediately that breastfeeding was not going well. "[Diane] would scream every time I tried to nurse her and took to sleeping for longer and longer periods," she observed. When wet diapers became fewer and farther apart, Marie took action and brought her then six-pound baby to the doctor. "He instructed me to mash a small piece of ripe banana, gradually increasing it to one-half banana by the end of a week, and add it to a mixture of one and one-half teaspoons honey to one pint boiled water, plus whatever breast milk I could express," she reported. Baby Diane "readily took the banana-honey mixture," by spoon, but continued to refuse her mother's attempts at breastfeeding. In the early 1960s Marie was among a small handful of women who were at least partially educated (usually through their own efforts to read widely) in some of the most basic principles of breastfeeding. Marie also found her way to the growing organization of breastfeeding mothers, La Leche League, which had sprung up just a few years before. It was with the help of League members that Marie obtained the use of an electric breast pump from a nearby mother. With the hope of stimulating her milk supply, Marie set about pumping and continued to put her baby to her breast in an attempt to get her to latch on. Within a week, with the assistance of League mothers and her electric pump, Diane went from her emergency banana diet to breastfeeding nine times a day. At the end of the week, her doctor "advised eliminating the honey-banana mixture," and Diane continued her life as a breastfed baby. Marie relied on the pump at first to simply help express larger amounts of milk to feed her daughter the banana mixture, but within a few days found that using the pump for just a few minutes before she tried to nurse her baby "started the flow of the milk and made the nipple easier for the baby to grasp."[1] This happy success story appeared in the *La Leche League News*, the group's member newsletter. In the article, Marie praised not only the League for its help and support, but the electric breast pump, which she stopped using once her "faith in the breast [had] been completely restored."[2] Given the appearance of this story on the front page of La Leche's newsletter, one might assume that unabashed breastfeeding advocates fell naturally into line as champions of the development and use of breast pump technology. In fact, the nature of the relationship between this technology, mothers, and the breastfeeding community more broadly has been difficult to pin down. At times, the dissonance between the breast pump and natural motherhood has dominated, while at others—as in the case of Marie Lawe—the breast pump helped mothers continue to breastfeed "naturally" when nothing else could.

In the 1940s and 1950s, electric breast pumps fell without controversy under the classification of a medical technology. The idea that pumps would move beyond the walls of hospitals and clinics in any great number had been batted about earlier in the century, but breast pumps, even manual ones, remained relatively novel and absent from the experiences of those who pursued breastfeeding until well after World War II.[3] For much of the twentieth century, the electric pump, in particular, remained the tool of the nurse in cases of postpartum breast issues in the hospital, such as engorgement and damaged nipples.[4] In a parallel use, pumps extended and modernized the "wet nurse" in American infant care by allowing physicians to control, manage, and monitor the collection of breast milk from donors without violating the taboos of twentieth-century bodily integrity, contagion, and intimacy and while keeping the scientific management of infant feeding in the hands of medical experts.[5] Given this history, then, it becomes far less surprising that by the time companies with names like Egnell, Medela, and Ameda introduced and marketed breast pumps widely in America (to a limited extent in the 1960s, and in a much larger way in the 1980s), breastfeeding advocates expressed concern over how to fit a device with a history as a technology of medical expertise into a construction of nursing an infant as a natural and "womanly art." Even as La Leche League served as a resource for mothers like Marie, who turned to breast pumps under dire circumstances as a means of continuing or salvaging breastfeeding, members ranged in their opinions about the widespread use of a technology that could also put milk into bottles and detach breast milk from breastfeeding, and from its ideological roots in natural motherhood.

Historians and others who have examined the breast pump have focused primarily on the question of whether the device should be embraced by women as a feminist technology, or viewed warily as yet another tool of female oppression.[6] My exploration of not only the long historical development of this technology but also its sharp rise in visibility and use at the end of the last century should make it apparent that scholars need not advocate one side or another. Much like historical debates over the use of pain-relief technologies during childbirth, the dominant construction of breast pump technology has changed wildly from time to time and place to place.[7] Furthermore, whatever the dominant construction of the breast pump has been at a certain time and place has consistently been challenged and contested by alternative voices, experiences, and viewpoints. The story of the breast pump provides the final strand in understanding the evolving relationship between the ideology of natural motherhood and breastfeeding at the end of the twentieth century.

The Origins of the Modern Breast Pump

The modern electric breast pump is rooted in a far less complex ancestor. In the nineteenth-century medical world, swellings, stoppages, and inflammations often required the extraction of some offending obstruction through the administration of a purgative or emetic, or even bloodletting or cupping. According to Audrey Davis and Toby Appel in *Bloodletting Instruments in the National Museum of History and Technology*, "mothers with underdeveloped or inflamed breasts posed a frequent problem for the nineteenth century physician."[8] Physicians, surgeons, and pharmacists of all sorts developed or modified their own handheld suction devices, the majority of which worked on the principle of creating a minor vacuum through the extraction of air by hand or mouth. A Philadelphia navy surgeon named Robert J. Dodd patented a device in 1844 that "consisted of a metal syringe provided with a plate of lancets that screwed on to a glass tube with a protuberance for collecting blood."[9] The pump was multipurpose, with the ability to be "adapted for extracting milk from the breasts of women by attaching a metal cap with a hole just large enough to accommodate the nipple."[10] These devices worked to assist healers in the extraction of fluids but could also be put to work to alleviate pain and inflammation due to breast engorgement and breast infections; to correct nipple inversions, which hindered successful latch-on; or even on occasion to "collect the milk so that it could be fed to the infant."[11]

These early devices were often designed to allow self-use "for reasons of modesty" in a period when a male physician treating women's health issues was still somewhat taboo.[12] Because "cuppers" and surgeons designed these early pumps for the purposes of medical extraction of blood and milk, they did not always include a successful or convenient system for capturing and storing the milk so that it could be fed to an infant. Those that did generally utilized the addition of "a glass protuberance" that could be "exhausted by syringes or rubber bulbs" in order to collect the breast milk.[13] A late nineteenth-century treatise written for mothers suggests, however, that by the 1890s manual breast pumps were a familiar device to be implemented in cases of "distention of the breasts" (or engorgement), though not before the sufferer had tried first to have the infant, then a nurse, or "an older and stronger child" suckle at the breasts. Even then, the author suggested a "young pup" would be as good a substitute as a pump if none was at hand.[14]

As breastfeeding rates fell dramatically between the late nineteenth century and the 1920s, and hospital births became more and more normalized, medical problems due to un-nursed breasts or poor breastfeeding arose as routine occurrences in postpartum care. That breast problems arose frequently throughout the period when breastfeeding was declining and hospital births increasing is further reflected in the observation by Davis and Appel that "from 1834 to 1975, more than 60 breast pumps were patented, the majority in the period from 1860 to 1920."[15] By the 1920s, the prospect of building an electric pump that required less manual labor became a reality. In 1923, dairy machine engineer, inventor, and international chess champion Edward Lasker filed for a patent on an electric breast pump that was designed for use in "maternity hospitals."[16] Working in partnership with Isaac A. Abt, the famous

Maw, Son and Thompson, "Woman Cupping Her Breast," *Surgeon's Instruments, etc.* (London, 1882). (The Smithsonian Institution Libraries, National Museum of American History, SI photo 76-13540.)

American pediatrician, Lasker brought his engineering know-how and experience working with dairy machines while Abt framed the problem and helped test out the devices.[17] Abt, himself a longtime proponent of breastfeeding, had spent the better part of two decades thinking about a way to prevent babies from suffering "the effects of artificial feeding."[18] "The result of our work," Abt would later proudly recall, "was an intermittent pumping device, powered by a small electric motor that could be connected to any lighting outlet."[19]

Nurses quickly became the most common users of the new technology and they responded enthusiastically to the device. One nurse writing in the *American Journal of Nursing* praised its efficiency "in drawing out inverted or flat nipples" and its ability to empty "the breast more completely and readily than does any other method we have used."[20] Wet nurses at the Sarah Morris Hospital utilized the pump for two years with

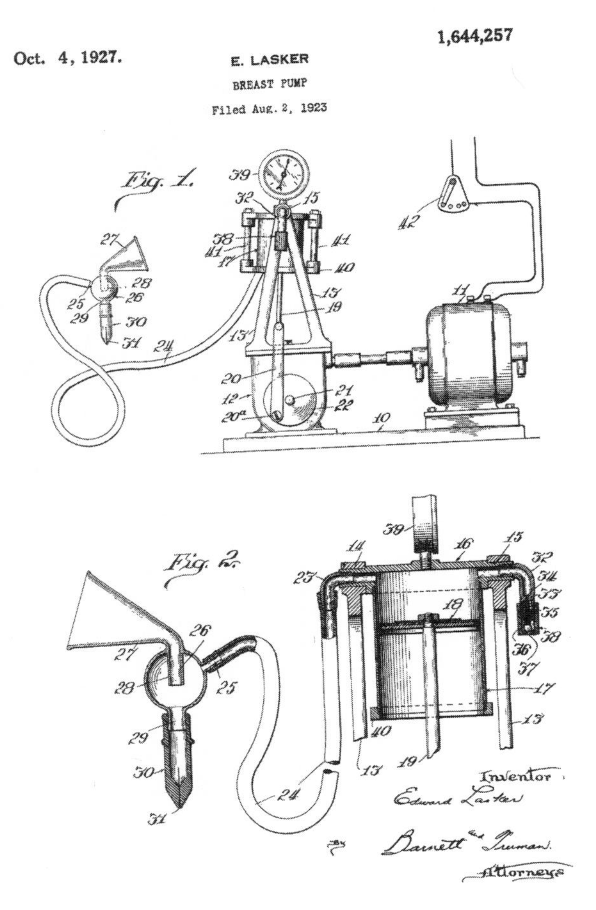

Technical drawing of Edward Lasker's pump, also known as "Abt's Pump." (Patent No. 1,644,257, "Breast Pump," United States Patent Office, issued to Edward Lasker of Chicago, Illinois, 4 October 1927. Application filed 2 August 1923.)

good results, and with the cooperation of the Chicago Lying-In Hospital, Lasker and Abt confirmed that "the pump operated satisfactorily and proved to be very useful [for] stimulating the secretion of milk and maintaining the supply for premature, weak, or sick babies."[21] The device also "met with rather general approval" from physicians, according to Abt, who demonstrated the device at the next American Pediatric Society meeting. While Abt and Lasker developed the technology for a specific medical population (i.e., sick or premature babies with poor sucking reflexes), the emergence of electric breast pump technology invariably led to questions about when to use it, who would use it, and why.

There were some in the medical community, for example, like Los Angeles physician Earl Tarr, who looked upon the arrival of the breast pump with a prescient apprehension about the impact of the device on medical oversight of infant feeding. Writing in 1925, Tarr argued that the pump ought to become a standard part of a physician's tool kit for better managing the breastfeeding mother and her infant. He believed there was no such thing as a "new-born infant [who] is physically able, during the first few weeks of life, to empty a breast." "The electric breast pump," Tarr argued, "can be used by [the doctor] to wonderful advantage" in the "the maternity division of the hospital," establishing good milk production in all mothers before they returned home—something he believed the majority of infants, healthy or otherwise, were often incapable of doing their own.[22] Tarr explicitly argued against the prospect of sending mothers home with these devices, believing that they held far more promise in the hands of the physician than in those of the mother. Tarr's concerns were not entirely isolated. In his memoirs Lasker himself recalled that the famous pediatrician and textbook author Joseph B. DeLee wrote to tell him that he "considered the machine indispensable in any hospital in which maternity work was done," but made no mention of the device as a domestic technology.[23]

By designing electric breast pumps, inventors like Lasker attempted to usher the age-old device into the twentieth century, creating a space for the pump inside the walls of the technologically modern hospital.[24] Until World War II, the United States served as the world's primary manufacturer of these pumps, with evidence suggesting that, in Sweden at least, the "Abt pump" was the dominant pump. When war broke out in Europe in 1939, restrictions on inter-Atlantic trade left some hospitals without a supply of replacement pumps. It was that year that a Swedish engineer named Einar Egnell, whose early endeavors to build war planes, cars, and steam engines had garnered him little financial success, learned about the expanding need for better breast pumps from Dr. Willers, a Swedish obstetrician whom he reportedly met at a dinner party.[25] Intrigued by the prospect of building a better breast pump for Willers's patients, Egnell embarked on three years of research studying the mechanics of lactation and milk extraction. He employed the assistance of Maja Kindberg, the head nurse of Stockholm's Södersjukhuset Hospital. With Kindberg's guidance and the participation of many new mothers in the hospital's maternity ward, Egnell designed the first breast pump with a built-in mechanism that successfully mimicked the sucking of an infant at the breast. Rather than providing continuous suction, Egnell implemented a three-period cycle that included two "suction periods" followed by a "resting period" into his pump in order to reduce the pain, nipple fatigue, and monotony that women often experienced when exposed to the continuous suction of the vacuum in the Abt pump. Nurse Kindberg, or Sister Maja, as she was known, advised Egnell on ways to make the pump more acceptable to new mothers. According to Egnell's family's records, upon bringing the first completed model of the pump to the hospital in 1941, Kindberg rejected it on the grounds that the large pressure gauge on the device would "disturb the patients."[26] Soon after making more user-friendly modifications, mothers at Södersjukhuset began to demand what became called the Sister Maja Breast pump (or SMB pump) over the existing Abt pumps, citing it as being more comfortable.[27]

In his research on human lactation, Egnell also helped invert the idea of a passive breast acted upon by an active infant (or pump, or hand, etc.). "The fact is," he stated, "that if there were no pressure in the alveoli to push out the milk no child or pump could get out a drop."[28] Egnell's research provided a scientific explanation for the observation that many nursing mothers had no doubt made from personal experience: breast milk is ejected from the breast, not pulled out. Egnell's observation that the amount of milk that could flow out of a breast was both temporarily finite and governed by a specific pressure differential that originated in the breast itself led to the creation of a pump with a lower suction rate than previous pumps. This was a significant improvement for mothers, as Egnell recorded a 33 percent rate of skin rupture with use of the Abt pump. Egnell's approach to breast pump technology resulted in a pump that not only worked better than other versions but actually appeared to be *more natural*—more like a nursing baby—than other methods of obtaining breast milk, more so even than hand expression. Built in conversation with his users, the mothers and nurses themselves, the SMB pump was designed to be the closest approximation of a suckling infant that a human-built machine could offer. While hand expression was laborious and, as Egnell suggested, "brutal," whether performed by mothers themselves or nurses, his breast pump operated far more "naturally," and like an infant; its use could effectively stimulate milk production in a situation where the infant could not latch on well enough to do so. This meant that while the pump could still be used to treat engorgement, it could legitimately improve a mother's chances of successfully breastfeeding her infant.

While Egnell's pump slowly grew in popularity in hospitals across Europe, it remained largely unknown in the United States until it slowly entered the American experience in the 1960s. In 1965, the Egnell pump caught the attention of none other than Niles Newton while she was visiting England. In 1965 Newton wrote to the League secretary, Edwina Froehlich, to share news that during a hospital tour she had learned "of an excellent new breast pump which was superior to any used before." The secret to its superiority, she wrote, was that "it not only sucks [but] it then lets go with a push." Such a subtle design change led, in Newton's opinion, to an experience "*more like natural suckling*" (emphasis mine) than other pumps she had seen. Newton even suggested she might purchase one herself for the League.[29] Enclosed in her letter she included some promotional material for "Egnell's Breast Pump," which acknowledged its utility as a rental unit that the mother "can conveniently use . . . in her home." The primary selling points, however, stressed its medical utility, including for use with babies who were unable to suck "in cases of harelip or prematurity," "when the mother's nipples are inverted," when the mother suffered from "cracks in the nipples," and "when breast feeding has to be suspended temporarily because of the mother or the child becoming ill." Still, the sales literature even in this early period hinted at the technology's possibility for expansion to a much broader user base when it suggested that the pump could be used for mothers who suffered from "hypogalactia" or "too little milk" as well as in cases when the mother "has more milk than the baby can use."[30] Newton's recommendation of the pump sparked the La Leche League to write immediately to the company and inquire about obtaining one.[31]

Egnell sold his pump on the grounds that it was "more natural" than many of the alternatives, an idea that caught the attention of Newton, La Leche League, and countless others. Egnell also pitched his device beyond the traditional medical market when he suggested any mother who produced too much or too little milk could use it. In constructing a breast pump that worked "naturally" to aid mothers with a broad spectrum of subjective breastfeeding issues, Egnell in effect helped broaden the pump's impact and desirability. Even before the SMB pump became a standard in the U.S. market, however, American mothers had already been slowly warming up to the idea of integrating certain technologies

Portrait of Niles Newton, photographed by Brenda Black. (La Leche League International Records, Special Collections and Archives, DePaul University Library, Chicago, Illinois.)

into their definitions of breastfeeding. Early forms of these technological nursing aids bore little resemblance to devices of masculine medical authority, like hospital-based breast pumps, which mothers in La Leche League sought to avoid.[32] Instead, League members and their allies traded information on the construction of homemade and locally made nursing bras, shirts, dresses, and even baby slings, in order to facilitate the integration of breastfeeding into their daily lives.[33] Figuring out ways to nurse discreetly troubled even the most defiant breastfeeding mom in the 1950s and 1960s.[34] Actively developing, modifying, and trading insider knowledge about these technologies in response to this culturally created "problem," which hindered the freedom and flexibility of a nursing mother, seemed, in fact, to be the very opposite of the passive utilization of the large, loud, and expert-operated breast pumps. Nipple shields, another technology that the breastfeeding community accepted early on, offer another example of this conceptual marriage between nature and technology that began in the 1950s. Nipple shields could be useful for the "correction" of inverted and flat nipples, in cases of extremely sore nipples, or even in cases of sucking difficulty in the infant.[35] What all of these devices have in common, in addition to helping women manage their breastfeeding difficulties, is that they were simple in their construction and familiar in their packaging—nursing garments were disseminated as sewing patterns, and nipple shields could be purchased easily at a local pharmacy or by mail order. When mothers implemented these technologies, they actively expanded the body of tacit lay knowledge that made breastfeeding work in their modern lives. Furthermore, by integrating these devices into their conception of natural motherhood and breastfeeding, the breastfeeding community that began to emerge in the 1950s laid the groundwork for a concept of breastfeeding that could be seen as completely "natural" while still incorporating certain technologies.[36]

Breast pump technology in the postwar years, however, did not fit easily within this carefully balanced construction of breastfeeding under the ideology of natural motherhood. By physically intervening in the mother-child relationship that "natural" breastfeeding required and by allowing mothers the flexibility of bottle-feeding without risking the hazards of formula, breast pump technology suggested what was for many an unnerving possibility, that of *un*naturally breastfeeding. The addition of a machine into the breastfeeding pair challenged the ideology of the natural when it posited the modern specter of a cyborg mother, embedded not with the natural, but with the decidedly unnatural form of the machine.[37] Perhaps most immediately threatening to many breastfeeding advocates was the disembodiment of mother's milk. The modern breast pump seemed to call into question the central importance of the mother-infant pair that natural motherhood relied upon. With the advancement of electric breast pumps it eventually became conceivable that a baby could actually obtain everything it needed by drinking it out of a bottle. Even more distressing to many breastfeeding and children's advocates, however, was the growing threat that as women entered the workforce in greater and greater numbers, they would leave their children behind with bottles of breast milk.[38]

Despite the fact that the design of the device became more and more sensitive to the physiology and psychology of lactation over time, its increasingly complex construction effectively shut mothers out of the process of technological production and positioned them solely as consumers. In looking back over the history of the modern pump we see a much longer trend: inventors of the pumps were overwhelmingly (if not entirely) male. One might almost imagine that if men had required the use of breast pumps for themselves that they would have been sold in kits, necessitated assembly, and allowed for personalization and enhancement with the purchase of additional parts and the ingenious tinkering of its user.[39] This was not the case, however, for breast pumps or their female users. The electric breast pump, therefore, while increasingly available as a consumer device, never comfortably settled in beside the homemade technology of the nursing bra or the simple and inexpensive nipple shield.

By 1970, La Leche League received regular requests for information and recommendations on breast pumps. Yet, despite the fact that they continued to help women temporarily access expensive, electric pumps as needed, the League remained officially and somewhat puzzlingly silent on the issue of the pump. As local League leader Judy Torgas explained in 1970, when confronted by mothers with questions about pumping, her standard reply was that "the baby is the very best 'pump' and we really aren't experts on any other kind!"[40] By that time, however, it was possible in many parts of the country for a mother to rent an electric pump herself, either through a hospital, her doctor's office, or even a local breastfeeding support group. Torgas herself acknowledged that this rise in breast pump use seemed inevitable, because "as more mothers are realizing the advantages of breastfeeding, more of them are determined to continue despite emergency situations where weaning might have seemed inevitable."[41] By the end of the 1970s, a growing segment of the breastfeeding advocacy community was beginning to acknowledge, albeit grudgingly, the potentially important role that breast pumping would play in expanding breastfeeding to working mothers and their babies.[42]

The decade of the 1980s would, in fact, witness the emergence of an entirely new profession, lactation consultants, many of whom would rely heavily on the breast pump (and other technologies) as a trademark tool. As breastfeeding grew in popularity, more and more mothers initiated breastfeeding and increasingly demanded the right to provide breast milk for their children regardless of their socioeconomic or employment status. In this context, many began to argue that the breast pump, once a technology symbolic of medical authority over infant feeding and maternal processes, could be seen as an incredibly important feminist technology. The debate over the breast pump, first framed in the early days of the breastfeeding movement and steeped in the ideology of natural motherhood, would become much more heated in the final decades of the century. Wrapped up in the midcentury ideology of natural motherhood, breastfeeding underwent yet another important reconstruction in the 1980s and 1990s as the movement fractured into increasingly divergent perspectives on the relationship between nature, technology, and motherhood.

The Expansion and Medicalization of Breastfeeding in the 1980s and 1990s

The year 1985 was a big one in the history of breastfeeding. It was the year in which 196 lactation consultants gathered in Washington, DC, for the first official meeting of the International Lactation Consultant Association (ILCA), the first such association for those who worked in lactation support and education.[43] Pulled together the prior year by a small group of La Leche League alums, the organization provided a tangible component to the gains of the movement's long struggle.[44] Its early membership and leadership, however, also reflected the growing influence of nurses and other medical professionals in the breastfeeding movement. As one of the founders, Kathleen Auerbach, acknowledged, "LLLI was instrumental in assisting this new allied health care field to become established."[45] "Many of us," she later reflected, "learned much about breastfeeding through our volunteer work as members and leaders in community-based breastfeeding groups, the largest and most visible of which is La Leche League."[46] Linda J. Smith, a fellow founding member of ILCA, agreed that the origins of the association began in 1982, "when La Leche League formed the Lactation Consultant Department."[47] Smith became the assistant to the department's director (JoAnne Scott) and "over the next two and a half years," she and Scott spent their days "researching credentialing, exams, educational programs, attending meetings on professional regulation."[48] The two eventually contacted Faith Bedford, a lactation consultant at the University of Virginia Hospital in Charlottesville, and asked if she would take up the cause of forming a professional association for lactation consultants.[49] The end result was the formation of ILCA, an organization that its leaders hoped "would unify all LCs and those interested in the field into a cohesive body large enough to make an impact."[50]

Simultaneously, La Leche League International was also helping to launch the International Board of Lactation Consultant Examiners (IBLCE), which administered its first credentialing exam to 248 candidates in Washington, DC (and 12 in Melbourne, Australia) in 1985. Prior to the formation of the IBLCE, people referred to themselves as lactation consultants but they were left to build their professional roles with little guidance, patching together credentials from regional, state, and local programs when possible and often building on an educational background in nursing (i.e., RN, BSN, MSN) and/or involvement with La Leche League or other such volunteer groups. The most well known and widely utilized of these pre-IBLCE programs was the University of California at Los Angeles Extension Program, which offered college credits in its "Lactation educator training program" and a certificate in lactation education (CLE).[51] For some time, CLE certification competed to serve as the field's gold standard, though it has largely given way to the International Board-Certified Lactation Consultant (IBCLC).

What had begun as a small group of volunteers in 1956 grew by the early 1980s into one part of an expanding professional career path. Biographies of the candidates for some of the organization's first officers reveal that both ILCA and La Leche League often drew from a common pool of breastfeeding experts. Faith Bedford, a key founder of ILCA and the organization's first president, had worked as a paid lactation consultant for eight years at the University of Virginia Hospital, where she educated nursing and medical students, interns, and residents about breastfeeding. Other candidates included longtime members and volunteers from the Childbirth Education League, the Nursing Mothers Council, and La Leche League International.[52] In her first presidential address in 1985, Bedford praised the connections that LCs had with lay groups as "vital and important to the continued growth and strength" of both lay groups and professionals.[53] She stressed common goals and core values, stating, "Lay helper, RN, clinic- or hospital-based, LC or student: we all share a need for a strong, unified association where, despite our different backgrounds, we can come together."[54] The connections stretched beyond the groups' leadership: of the 259 people who sat for the first IBLCE examination, 133 were La Leche League leaders and 93 were registered nurses.[55] Yet, despite the close personal and organizational ties between La Leche League and the newly emerging profession, territorialism quickly acted to distance ILCA's professionals from LLLI's lay people

Lactation consultants, in their own quest to be recognized as legitimate professionals, sought to distinguish themselves from their volunteer roots, to be taken seriously as colleagues by physicians and nurses, and to untangle themselves from the organizational apron strings of groups like LLLI with its rigid interpretation of natural motherhood ideology. As they moved into medical settings, they also moved further and further from the living rooms and kitchens where it had all begun. Instead, they sought recognition and payment, from doctors, nurses, and mothers who would come to them not for a life lesson or a personal growth experience, but for help with a specific breastfeeding issue. Unlike their League predecessors, lactation consultants were not in the business of spreading the gospel of natural motherhood, but of helping lactating mothers overcome specific breastfeeding obstacles. In creating an autonomous profession, lactation consultants worked to expand lactation support to more mothers than ever before, while at the same time they distanced themselves from the radical implications of natural motherhood ideology.

The inevitable professional growing pains that ILCA experienced often played out quite visibly in the association's journals and newsletters. Over the 1980s and 1990s, the breastfeeding community, once confined to a handful of organizations, found itself on multiple and increasingly divergent paths: on the one hand, community-based volunteer organizations like La Leche League International continued to align with a more holistic and ecological view of breastfeeding and a model of experiential knowledge transfer based on intimate mother-to-mother interactions. On the other hand, professional lactation consultants set themselves apart from their volunteer pasts when they sought compensation for their services, turned to increasingly specialized scientific knowledge about breastfeeding as a means of credibility, and developed expert medical knowledge about the pathologies of the nursing pair. Lactation support workers also debated what kinds of credentials, education, and experience should be required of someone in order to legitimately claim the title of "lactation consultant." After attending her first ILCA annual meeting, Jeanine Klaus of Tennessee put her summation of the situation in blunt terms: "I was not prepared for those who raised their flags from the pinnacle of their achievements, shouting, 'Follow me, it's the only way!'"[56] At the root of these divergent paths sat tensions over the ideological ties between breastfeeding culture and natural motherhood and the different visions of what the future of motherhood and breastfeeding should look like. While the professionalizing turmoil of a new health care field included the expected disputes over credentialing and education, the growing split between lay and professional breastfeeding supporters often came to a head in discussions over the relationship between technology and breastfeeding.

In the earliest volumes and issues of the association's journal, the rhetoric and tone of the editorials and letters returned time and time again to the natural motherhood ideology that had infused the lay breastfeeding movement. A 1986 guest editorial by RN, BSN, and IBCLC Chris Mulford from Swarthmore, Pennsylvania, for example, demonstrates the continuing resonance of the connections between "the natural" and breastfeeding:

> When a woman struggles to overcome obstacles . . . that interfere with her instinctive drive to protect and nurture her infant, she can tap into a wellspring of power at that deep level. Thanks to the Women's Movement, the association of power with the female sex is not an unfamiliar idea. But the power I'm speaking of doesn't have to be wrested from men. . . . Its source lies within women's life experiences. Rather than trying to force men to share their power with us, we can bring our own power to bear along with theirs as we confront the problems that face our human family. Breastfeeding empowers women.[57]

Mulford's perspective offered in second-wave terms what breastfeeding advocates had been claiming for decades. Her ideological stance alongside her training as a nurse and her credentials as an IBCLC suggest that, much like the women's reproductive health activists involved in abortion services and labor and delivery reform, many LCs saw themselves as bringing a female-centric perspective to the medical complex's engagement with "natural" bodily process.[58] Several of Kathleen Auerbach's early editorials reiterated a similar understanding of the LC's professional responsibility to continue carrying the mantle of natural motherhood, seeking to empower women through breastfeeding assistance and a connection to their instinctive selves. In one piece, Auerbach railed against any disunity in the breastfeeding community when she called out LCs whom she saw blaming women "for their own failure." In cases of "breastfeeding failure," she wrote, "the explanations suggest that the persons least likely to control hospital routines, the persons most likely to feel powerless in the face of authority figures, the people most in need of assistance and least likely to obtain it are to blame for their own difficulties."[59] "Even more disturbing," she noted, "is that this explanation is most likely to be offered by other women, some of whom have themselves experienced what it is like to be a hospitalized postpartum patient, a woman being looked down on (literally!) by authority figures in white clustered around her bed." For Auerbach, this situation was unacceptable. "We must stand together," she argued, "if we are to change societal attitudes, the routines of hospitals, clinics, other settings."[60]

Ironically, as the profession confronted the challenges of upholding its responsibilities as health workers and educators regardless of personal beliefs, LCs began to adopt a much more medical and technological model of breastfeeding. Although the effort to make holistic breastfeeding support available to more mothers had initially spurred their move to professionalize, LCs as a professional body eventually turned away from their ideological roots in order to claim greater authority and legitimacy. The shift was not (and most likely is not) complete nor has it gone uncontested, but as the *Journal of Human Lactation* (*JHL*) refined its goals and its relationship to the field under the editorship of Auerbach, natural motherhood ideology became less and less visible in the articles, editorials, and letters on scientific studies; malpractice insurance; and educational reform. In the same editorial in which Auerbach scolded lactation consultants for turning against women and being "their own worst enemies," she argued that if LCs wanted to succeed in their professional goals they had to do away with "the antagonism of LCs to women whose lives differ from their own." In making this statement, Auerbach explicitly addressed the long-standing discomfort in the natural motherhood community over the issue of "employed women." "According to a recent survey," she wrote, "breastfeeding initiation is actually higher among employed women than among mothers who are staying at home." Given the increasing rates of women who were fulfilling the dual roles of a "homemaker and a breadwinner," Auerbach argued that it mattered little what LCs thought personally about whether a mother should work. "Unless LCs clearly stand for all women, in all circumstances," she argued, the profession would do little to address the actual needs of those mothers who, in reality, needed their guidance the most.[61]

The obligation of LCs as professionals who were seeking to make a living with their specialized knowledge invariably required that they set aside their beliefs about the relationship of breastfeeding to the ideology of natural motherhood. Many LCs, particularly those who came to the field via a background in nursing or some other health profession, had few ideological strings to sever when they reached for the breast pump. In fact, some LCs found it professionally and personally lucrative to partner with breast pump companies from the start, acting as rental agents for pumps and other "breastfeeding supplies."[62] Such partnerships brought in more clients and referrals than many isolated LCs would have been able to bring in on their own. ILCA and the IBLCE maintained no rules about the relationship between breast pump manufacturers and lactation consultants, though they did actively prohibit them from working for any baby food or formula companies.[63] LCs routinely and proudly wrote in to the *JHL* to share stories and strategies for success and professional autonomy. In many of these self-made stories, the LCs' paths intersected in one way or another with breast pumps. In response to this trend, Kathleen Auerbach wrote an editorial asking the *JHL* readership to consider the fact that many breast pump companies included "artificial teats in the packaging of their pumps." She pressed further, suggesting that the lactation consultant who "helps mothers learn how to use such devices" may have to confront the possibility that she could be unwittingly "contributing to the practice of bottle-feeding."[64]

Lactation consultants did know that the early introduction of bottles or pacifiers could lead to the dreaded and highly contested phenomenon known as "nipple confusion." Though it remained controversial among physicians and nurses, breastfeeding advocates in both the lay and lactation consultant communities could provide plenty of scientific, clinical, and anecdotal evidence that babies who drank from bottles or soothed themselves on pacifiers often gave up on breastfeeding after essentially "forgetting" (or never learning in the first place) how to drink properly from the breast.[65] A typical LC client backstory, in fact, often went something like this: The baby had been kept in a nursery, separate from the mother for long periods during which nurses had remedied his or her cries in between scheduled feedings by administering pacifiers and perhaps supplements. The baby, as a result, would develop a bad latch, resulting in a painful experience for the mother. Serious nipple damage might occur and the infant would not extract enough milk from the breast to keep the mother's milk supply up.[66] The mother, desperate to get her infant to stay hydrated and gain weight, would pump—maybe even resort to formula—and feed from a bottle. Compared to the nightmarish wrestling match she had been experiencing with breastfeeding, the bottle would seem a simpler, if not miraculous alternative. When the mother finally found her way to a lactation consultant, the breast pump would seem a magical cure for an already dying breastfeeding relationship, the perfect tool for turning the situation around. When used devotedly, the pump could stimulate milk production while the LC worked to "re-imprint" the nipple-confused baby with "orthodontic nipples," supplemental nurser systems, nipple shields, and plenty of patience.[67]

For the most part LCs recognized that in spite of its importance in these circumstances, the pump could also become part of the problem by shifting the focus from breastfeeding as a process toward breast milk as a product. This subtle shift could then increase the likelihood that infants would end up sucking on man-made, rather than natural, nipples.[68] In their day-to-day practices, however, some LCs looked to the pump as a critical component in the difference between a breast-milk-fed and a formula-fed infant. Given the alternative of artificial milk, LCs and their clients turned to the pump as the *most natural alternative* to breastfeeding. The compromise that many LCs learned to make was to recommend the delay of any bottle until after the third or fourth, or even the sixth week. By that point, many hoped, the mother and baby would be comfortable enough with breastfeeding that the infant would not "forget" how to suckle at the breast.[69] The danger that such a strategy posed for the working mother, however, was the rearing of an infant who would become so accustomed to the breast that he would refuse to take a bottle. As one LC wrote, "I have seen a fair number of babies who actually do refuse a bottle altogether—will not take one from anybody, any time, any way. It's easy to say, 'Well, smart baby!' But for a mom who really does have to work, this can be extremely upsetting, especially if the sitter is calling several times a day to report that the baby is screaming and starving."[70] In these scenarios, it seemed inhumane (and impractical from a professional standpoint) to do anything but offer a mother advice about pumping and bottle-feeding. LCs often advocated alternatives such as feeding with an eyedropper, syringe, spoon, or cup as "simple" and "easy," but the fact that these methods never really caught on suggests otherwise.[71] As one LC put it, "I'm becoming far less dogmatic about things [i.e., bottles]. I know I've 'lost' mothers because they just couldn't cope with cup feeding or using a syringe and I'd scared them off bottles (because I worried using the artificial nipple might create more problems and make things worse)." She explained further, "I'm the only show in town. . . . I do the best I can to keep breastfeeding going. While

I will never casually suggest using the bottle or a nipple shield, I now do so when I see that without something easier, the mother is just not going to make it."[72]

This was not the way all LCs positioned themselves in this debate, but it was increasingly the way that those affiliated with ILCA and the IBLCE did. In 1991, U.K. breastfeeding advocate, counselor, and activist Gabrielle Palmer published a sharp critique of the technological and medical direction in which she saw the field going: "In our keenness to raise the status of breastfeeding expertise, let us not fall into the trap of inadvertently signaling to mothers that breastfeeding is a matter that requires mediation by experts. Our goal should be to prevent those problems caused by medical misdirection and commercial and social pressures and ultimately do ourselves out of our jobs."[73]

In raising the possibility that lactation consultants themselves might contribute to the medicalization of breastfeeding and "the infiltration of commercial interest," Palmer called out the breast pump as a primary source of concern. "I feel worried about the direction of lactation consultancy when I read on professional cards the words 'breast pumps' or 'breastfeeding equipment,'" she wrote. Such a conflation of lactation consultancy with this equipment, she feared, "may indicate to a new mother that technology and devices are an integral part of the process." "Breast pumping is not breastfeeding," she argued, and while it "may be a necessary second-best," for many mothers, such a situation was not—nor should it be—the goal of the lactation consultant. "The mother working outside the home, whose commitment to her child is demonstrated through her use of pumps, is an oppressed woman," she argued, before pointing out the ironic parallels she saw developing between the history of doctors and formula companies and lactation consultants and pump manufacturers. "It would be disastrous if the zeal to gain credibility as professionals . . . were to mean that lactation consultants fell into the same trap that ensnared doctors and become beholden to pump and device manufacturers."[74]

Not surprisingly, Palmer's provocative essay elicited passionate responses from the *JHL*'s readership, including one from Kathleen Auerbach herself, who followed up Palmer's piece later that year with her own appraisal of the situation. "The number of articles about lactation appearing in the professional literature has increased dramatically in the last two decades," she observed, noting, "if one categorizes these articles by topic, it is clear that the larger group focuses on breastmilk as a product."[75] Though Auerbach refrained from a direct critique of the profession she had helped found, her sentiments mirrored those of Palmer and drew upon natural motherhood ideology to distinguish breast milk from breastfeeding. "The medical model supports thinking in terms of product: its adequacy or inadequacy, its availability, and the degree to which it supports and sustains an adequate nutritional outcome. . . . The view of breastfeeding as a process forces the investigator to consider the human quality, the interaction that sustains production and renders its transmittal from producer to user such an intimate experience, and one that cannot be duplicated with artificial substitutes."[76] Both Palmer and Auerbach overtly linked the use of breastfeeding technologies with a process of medicalization and the "destruction of knowledge that was once common to all."[77] Such a perspective reflected a strong connection to the breastfeeding movement's roots in natural motherhood, with its emphasis on female systems of knowledge exchange and technologies controlled and often handmade by women for women.

Letters to the editor in response to this ongoing debate revealed that within the lactation support community these concerns continued to percolate. One IBCLC from Atlanta wrote that she had been "excited and relieved" to read Palmer's essay "because hers was the letter I have been composing in my mind for the past four years."[78] "Over the fifteen years that I have been working with breastfeeding families," she wrote, "I have been increasingly dismayed at the growing emphasis on gadgetry, pathology, and the medical model as applied to a function that is ultimately the province of mothers." "We as lactation consultants," she argued "must guard against interfering with the norm instead of simply supporting it."[79] Others also wrote in to share their appreciation for Palmer's staunch views, including Judy Hopkinson, PhD, from Baylor College of Medicine in Texas. "I too have been alarmed at the growing liaison between breast pump manufacturers and lactation consultants," she observed. Hopkinson feared that, if not checked, the continuing expansion of the breast pump as a consumer device would lead to the further "oppression of women and the devaluation of motherhood."[80]

Not all LCs agreed with these critiques, including those who allied themselves with breast pump companies and who worked with mothers who relied heavily on the devices. Debates over this issue often took on a personal and heated nature. IBCLC Alison Hazelbaker, for example, argued that "breast pumps are a routine part of keeping American mothers breastfeeding." She went on to add that she could not "see how renting an electric pump to the mother of a hospitalized, premature baby is a co-option, how selling a supplementer to the mother of an adoptive baby is gnostic cultism, or how using suck training to reimprint a nipple confused baby who refuses the breast makes me a member of the cognoscenti." "It is my job to see that mothers who ask for my help receive what I have to give," she stated, before adding definitively, "I will use whatever tools are at my disposal. . . . What works, works."[81] Many LCs seemed to share Hazelbaker's opinions, and the direction of the field consistently pointed toward breast milk feeding as a "type" of breastfeeding that was far preferable to any artificial alternatives, even if it was not as natural as feeding an infant at the breast itself. The *JHL* published information on the storage and handling of mother's milk and routinely featured articles, letters, and discussions in which authors treated the breast pump and other breastfeeding technologies as relatively harmless tools of the trade.[82]

WIC and the Expansion of Breast Pumping

Published arguments about the relationship between the pump and working motherhood often tended to imply a middle-class and white mother as the focal point of natural motherhood ideology. Conversely, in the discourse surrounding breastfeeding mothers in the Supplementary Food Program for Women, Infants, and Children (WIC), the pros and cons of breast pumps were rarely if ever discussed. With its enactment in 1974, WIC offices began assisting economically disadvantaged families through the provision of nutritional resources and education to pregnant and postpartum women and children.[83] Though the program had always received government support for breastfeeding mothers, by the end of the 1980s, WIC participants saw an increasingly organized and well-funded effort to promote breastfeeding by various means, including offering longer-term and enhanced nutrition support to women who breastfed their infants up to one year.[84] The passage of the Child Nutrition and WIC Reauthorization Act of 1989 required the U.S. Department of Agriculture (USDA) to promote breastfeeding and, even more importantly, set aside $8 million for breastfeeding "support activities," and for "the purchase of breastfeeding aids by WIC agencies."[85] The breastfeeding support community was overwhelmingly pleased by the act's passage, which organizations including LLLI, ILCA, and the

American Academy of Pediatrics helped bring into being as part of an initiative launched in 1988 and funded by the American Public Health Association (APHA): the National Committee for Improving Breastfeeding Strategies in the WIC Program.[86] In the wake of the act's passage, lactation consultants mobilized to take advantage professionally of the new demographic that was now being brought into the breastfeeding community. "Now the work begins!" wrote Nancy Schweers, the ILCA representative to the APHA committee and an IBCLC from San Antonio, Texas. Schweers encouraged ILCA members to lend their "experience and expertise" and to grab hold of this "unique opportunity and responsibility" that confronted them. "We all will learn a lot when we work to offer women on WIC the health benefits, dignity, and autonomy that breastfeeding provides to lactating women and their nursing infants," she wrote.[87] Long a movement dominated by the white middle class, the expansion of WIC's breastfeeding efforts promised to at long last bring poor and minority women into the flock. In expanding breastfeeding to more and more women from a wider range of backgrounds, experiences, and expectations, the efforts of WIC, ILCA, and other support groups stretched the process further and further from its ideologically bound roots. This stretching, however, allowed some women to maintain an experience of breastfeeding steeped in the ideology of natural motherhood, while others sought to obtain some of the benefits of breastfeeding through the use of modern technologies and markets.

It didn't take long for the impact of WIC's efforts among low-income mothers to trickle into the discussions of triumphs and trials in letters written to the *JHL*. By autumn of 1990, for example, Ruth Forni wrote to proudly share news of the opening of a new WIC-sponsored "Mothers' Room." "This beautifully decorated space," staffed by a WIC employee and a peer counselor, hoped Forni, would provide a room "for mothers to breastfeed their babies and to obtain help at the same site where they receive health care and obtain WIC food supplement vouchers." In addition to the space and lactation support, the program offered "breast pumps and a cloth baby sling" without charge, and without ideological conflict. "We hope that the breast pumps will encourage mothers to breastfeed even when they must be separated from their babies," wrote Forni. She also highlighted that by giving out cloth slings they were simultaneously encouraging "the mothers to maintain or increase body contact with their babies."[88] Paired with a baby sling, the epitome of a technology of natural motherhood, the breast pump appeared just as helpful, just as natural. In Forni's explanation, too, one detects the idea that the pump may facilitate distance between mother and baby, but the sling would act to counteract that effect. That both items were given out as valuable incentives to encourage a mother to breastfeed, however, also hints at the birth of a strong consumerist component to the breastfeeding culture of the 1990s. The idea that one "needed" any consumer goods to breastfeed, in fact, seemed to grow in tandem with the expansion of working mothers of young infants in the labor force.[89] As more women sought the experience of breastfeeding, the creation of a breastfeeding "market" promised increasing profits to companies who could sell products that resonated with the ideology of natural motherhood and breastfeeding.[90]

For some in the lactation support community, the pump not only represented the reach of the "male medical model" into the sphere of breastfeeding culture; it also represented the taint of mass-market consumerism in a process they saw as natural—and pure.[91] As Kathleen Auerbach shrewdly observed: "Breastfeeding mothers represent a potentially lucrative market for makers of . . . breast pads, breast creams and lotions, nursing bras, clothing, . . . formula supplements for those times 'when you can't nurse the baby,' vitamins, . . . breast pumps, nipple shields, breast shells, nursing supplementers, cushions on which to prop the nursing baby, and whatever other products are created with an eye toward reaching this group of women."[92] The impurity of the market threatened to contaminate the nature embodied in the process of breastfeeding without interference of gadgets or medical expertise. The consequences of this pollution could, feared Auerbach and others who shared this view, diminish or even negate the potentially empowering effects for mothers who breastfed "naturally." Despite these fears, few could deny the utility of the breast pump when it came to increasing the number of mothers who would feed their children breast milk, whether from the breast or the bottle. Critiques of maintaining a "pure" and "natural" construction of breastfeeding became even less powerful in the case of WIC mothers. Women in less-than-ideal economic situations that required government assistance deserved as much opportunity to breastfeed as the middle-class suburban mother, and few LCs could have questioned the importance of a breast pump for a working-poor mother who wanted to breastfeed.

As a reflection of this growing acceptance of a more flexible ideological connection between breastfeeding and natural motherhood, the

JHL's interest in WIC programs expanded over the course of the 1990s. In 1992 and 1993, for example, the journal featured articles on breastfeeding programs at WIC sites in Indiana, Florida, Utah, Alabama, California, and New Jersey, covering a wide spectrum of clients from urban, rural, racial, and ethnic backgrounds.[93] A common denominator across these sites was the provision of breast pumps. Some WIC offices also offered "a pumping station to enable mothers to use an electric pump in the office."[94] The Harrison County WIC office in southern Indiana went even further, helping to cultivate relationships between its breastfeeding clients and the local La Leche League group, providing transportation to and from meetings. The Harrison County program, in addition to the dissemination of breast pumps, also provided all its breastfed infants with T-shirts that proudly proclaimed "I'm Fed the Natural Way."[95]

In 1995, Kathleen Auerbach teamed up with Kathleen Bruce, a Vermont nurse, La Leche League leader, and International Board Certified Lactation Consultant (IBCLC) to start LACTNET. The Listserv created a virtual meeting ground for people working to support breastfeeding from around the globe.[96] Described as a "forum for cutting-edge information, discussion and support focusing on best practices, emerging thoughts and current research," LACTNET quickly became home not only to certified lactation consultants, but to lay counselors, nurses, doctors, midwives, public health advocates, writers, scientists, dietitians and doulas, and any one dedicated to supporting mothers who breastfeed.[97] The nature of the dialogue surrounding breast pumps and the professionalization of lactation consultants on LACTNET in the 1990s suggest that the tensions of the 1980s had largely dissipated, at least within the LC community. Overall, the question of whether breast pumps should be advocated by LCs seemed largely to be a thing of the past. By the mid-1990s, breast pumps kept women breastfeeding, or at the very least, breast milk feeding, when they might otherwise quit, and breast pump companies gave LCs jobs and clients. As the demand for breast pumps increased, so too did the supply; a variety of pumps flooded the market, primarily from the two largest pump suppliers, Ameda and Medela (both of which had their roots in Einar Egnell's original SMB pump), and a few newcomers like Bailey Medical (which sold the Nurture III and Double-Up pumps).[98] Given the growth in pump options, LCs increasingly saw a need for their services as consumer educators and advocates, in addition to, if not in lieu of, breastfeeding counselors. One LC located in Illinois offered the following list of services to potential clients: "demonstration of set up and usage, free consultation services regarding the pump, . . . collection & storage of milk, tips on pumping, working mother, and free replacement of any little parts they may loose [*sic*] [or] the cat gets hold of." "Whether we like it or not," she added, "we are in the marketing and sales world and we have to get our skills up if we are going to be competitive and not lose our services." In the twelve years she had been a practicing LC, she wrote, the field had "definitely changed."[99]

As the role of the breast pump expanded, the LC profession became inexorably linked to the technology. Though many LCs maintained a fundamental belief in the benefits of feeding an infant at the breast, they also saw time and time again how the improper application of the pump could result in just the opposite. As one LACTNET subscriber wrote:

> If I had been given a breastpump [*sic*] when I had left the hospital with instructions on how to use it, my first [baby] may have gotten breastfeed [*sic*] instead of being fed expressed milk from a bottle for [the] first six months. When my milk came in, my nipples went flat and [my baby] couldn't latch on. I called a LLL, who came right over to my house and tried very hard to be helpful, but she had never had to use a breastpump, and even though I had a good manual pump, neither one of us could figure out what we were doing wrong, and why we couldn't get it to work. By the time my brilliant husband figured out the problem, [my baby] had a strong preference for bottles and I never could . . . breastfeed.[100]

The bottom line was that not all pumps were created equal and managing pumping in a way that allowed feeding at the breast to continue (so that babies did not end up bottle fed) took care and expertise that, once again, most mothers did not *naturally* have. The lactation consultant increasingly became, whether she wanted to or not, an expert on breast pumping. LCs sometimes expressed overt frustration about the carelessness with which some mothers turned to breast pumps, disregarding or ignorant of the problems that unknowledgeable pumping could cause, including nipple confusion and latching issues. One LC recounted a tale of parents who wanted to implement breast pumping "for convenience" over the holidays. "Today [the] parents showed up, woman tearful, father defensive," she wrote. They "have decided to wean their 6 week-old because the baby now prefers bottles." The mother complained that the nursing and pumping schedule had left her breasts sore, the baby's "sucking pattern hurts," and she described the father as "grumpy about how much 'trouble' all this breastfeeding business is!"[101] Given these kinds of outcomes, the need for an expert who could help navigate the application of the quickly proliferating technology became undeniable, and LCs, increasingly based in clinic and hospital settings, were swallowed up in the medical-industrial complex of the American health care system.

Full Circle: Mothers, Milk, and Pediatricians

Pediatricians as a professional group and the American public health apparatus more generally arrived late to the breastfeeding party. When they did, however, they helped to create a more medicalized model of breastfeeding that focused on disembodied human milk rather than on the process itself. The efforts of medical professionals and policy advisers to support breast milk feeding alongside the expanding definition of "breastfeeding" to include "breast pumping" contributed to the widening gap between the construction of breastfeeding at the end of the twentieth century and its midcentury roots in natural motherhood ideology. The result was that by the turn of the twenty-first century, the definition of breastfeeding incorporated an ever-widening definition of experience, technology, and medical expertise that expanded the ability of more and more women to feed their babies using their own breast milk, while distancing mothers from the process of feeding an infant at the breast.

The slow mobilization of physicians and government resources for breastfeeding support at the end of the century arguably began in 1981. That was the year in which the American Academy of Pediatrics (AAP) established a "Task Force on the Promotion of Breast Feeding."[102] The assemblage of the task force came in the midst of a dramatic decade-long increase in breastfeeding rates and on the heels of the 1981 World Health Organization's Code of Marketing of Breast-Milk Substitutes. The efforts of the international health community to shape health policies on breastfeeding prompted a response from the academy.[103] While pediatric leaders endorsed certain attributes of the code, primarily its aim to "contribute to the provision of safe and adequate nutrition for infants by the protection and promotion of breast feeding," the task force took issue with what it claimed was the code's "narrow focus and essentially negative regulatory proposals" to reign in marketing practices of formula manufacturers—particularly in the developing world.[104] The task force convened to discuss how "physicians and others providing health care and counsel, hospital and workplace ambience, formula manufacturers and their marketing practices," influenced "mothers' choices of infant feeding practices." As a result of the discussions, the task force made recommendations in four general areas including improvements in pediatric education and training on the subject of breastfeeding, improvements in patient education, emphasis in hospitals on developing practices that "are supportive of breast-feeding," and improvements in the "environmental support system" for mothers once they have "chosen" their method of feeding. The 1982 statement did a number of things, including attempting to reposition pediatricians as experts on breastfeeding by producing a narrative of the AAP's history that made it seem as though the primary goal of pediatrics had always been to increase breastfeeding. In making its recommendations, it focused exclusively on what pediatricians as individuals should do, even while it acknowledged elsewhere that a complex set of factors, including hospital administrators and maternal support networks, also influenced breastfeeding. In short, the statement politically positioned the AAP in an enlivening global discourse on breastfeeding without setting any clear goals or making any radical calls for change at either the professional or social levels.

Two years later, in 1984, the U.S. surgeon general, C. Everett Koop, called the Workshop on Breastfeeding and Human Lactation to order. In the preface to the workshop proceedings, Koop noted that "research findings have documented the benefits of *human milk* [emphasis added] and lactation for babies and mothers."[105] Despite seeing a steady increase in breastfeeding throughout the 1970s and early 1980s, however, Koop also stated that the rise had occurred "predominantly among middle- and upper-income, educated, white women."[106] The workshop called together representatives from the health professions and voluntary organizations involved in breastfeeding promotion and support to help develop a plan to reach the goal of increasing breastfeeding to 75 percent of mothers at hospital discharge and 35 percent at six months postpartum. The workshop was chaired by Ruth A. Lawrence, a pediatrician from the University of Rochester Medical Center, now a well-known breastfeeding medical specialist and the author of numerous editions of breastfeeding textbooks for medical students and physicians.[107] Cochairperson Henry A. Thiede, also from the University of Rochester Medical

Center, represented obstetrics and gynecology. La Leche League (Betty Ann Countryman), the American College of Nurse-Midwives (Judith A. Flanagan), and the March of Dimes (Mary Hughes) were among those who participated, alongside the American Academy of Pediatrics, the American College of Obstetricians and Gynecologists (ACOG), and others.[108] Notably absent, of course, were the lactation consultants as a distinct group, because ILCA did not form until the following year, 1985.

The workshop participants ended the two-day meeting with a set of recommendations addressing six primary areas, many of which echoed the issues raised by the AAP in 1982: workplace issues, public education, professional education, the structure of the health care system, social support services, and scientific and medical research. Participants identified each of these as areas in need of improvement and support in order to increase breastfeeding rates across socioeconomic demographics. The third category, professional education, again focused explicitly on the lack of training available to medical practitioners in the subject of breastfeeding, stating, "It is imperative for all health care professionals to receive adequate didactic and clinical training in lactation and breastfeeding and to develop skills in patient education and the management of breastfeeding."[109] The critique of the health care system at large, however, was even more sweeping, stating simply that "the health-care system needs to be better informed and more clearly supportive of lactation and breastfeeding."[110] In their suggestions for addressing this matter, the workshop participants called for the following: the determination of national policy by the Division of Maternal and Child Health, which should convene a national group to advise and monitor that policy; the improvement in support for lactation in the Women, Infants, and Children (WIC) program; the development of policies and activities "which more clearly support breastfeeding" by professional organizations including AAP, ACOG, and the American Hospital Association, among others; the development of guidelines by the Joint Commission on Accreditation of Hospitals for hospital policies "which will promote fully informed choice about infant feeding and which will support a mother's decision to breastfeed (including rooming-in and feeding on demand); and the exploration of health insurance coverage for lactation counseling and breastfeeding support.[111] The workshop was an important and necessary step toward reshaping America's health system to better address the needs of breastfeeding mothers, and the recommendations helped influence the terms of the reauthorization of WIC in 1989.

In 1990, the Innocenti Declaration, organized by WHO and the United Nations Children's Fund (UNICEF), injected a growing sense of global urgency to American efforts to improve breastfeeding rates and duration. The declaration, made by WHO and UNICEF policymakers in Florence, Italy, on 1 August 1990, stated that mothers should exclusively breastfeed for the first four to six months, and "thereafter, children should continue to be breastfed while receiving appropriate and adequate complementary foods, for up to two years of age or beyond."[112] In response, that same year, at the request of the American Academy of Pediatrics, the USDA's Food and Nutrition Service co-launched, with the Department of Health and Human Services, the Breastfeeding Promotion Consortium. Made up of twenty-five organizations, including government agencies, professional associations, and lay groups, the BPC met several times, but by early 1993 had not yet made any formal recommendations.[113]

In 1994, the AAP's Advisory Committee to the Board on Community Pediatrics (ACBOCP) voiced concern over the fact that the BPC had not yet developed any promotion strategy and that the "formal meetings and activities of the BPC have been put on hold since May 1993." Later that year, however, the AAP balked when asked to endorse the strategy document that the BPC came out with—even though an AAP representative served on the committee. Again the AAP acted to gain a stronger foothold in the policy discussions taking place. Particular points of contention arose over whether or not the BPC's recommendations went too far in removing the pediatrician from conversations with parents about things like supplemental feeding and non-breastfeeding alternatives. ACBOCP argued that "parents should ultimately be the decision makers" in these matters. The AAP's Committee on Community Health Services (COCHS) also disapproved of what it believed was the low level of emphasis that the BPC's strategy placed "on the role of the pediatrician as an advocate to breastfeeding mothers." The result of these discussions was the recommendation by the ACBOCP that the AAP set up a Work Group on Breast-Feeding to be charged "with assisting the AAP on the refinement of its position on breastfeeding."[114]

As a result, the AAP moved again in an attempt to "take the lead" in the development of national health care policies surrounding the promotion of breastfeeding. This led for the first time in its history to the decision to cosponsor the La Leche League's annual Physician's Conference, meant to educate medical professionals about breastfeeding. In creating the Work Group, the board of directors tasked the members with developing an official AAP statement on breastfeeding that would clearly lay out their position on the promotion of breastfeeding, to replace the academy's 1982 statements.[115] Though it would take several years, the Work Group, consisting of ten members and headed by Lawrence Gartner, would go on to successfully author one of the most definitive and influential health care statements on the value of breastfeeding in recent American history.

Perhaps even more importantly, however, the Work Group not only successfully positioned the AAP as an authority on breastfeeding, but placed pediatricians firmly at the top of a solidifying scientific hierarchy in the management of breastfeeding. As part of this process of securing itself as an unquestioned breastfeeding ally, the AAP reached out to lactation consultants. In 1997 the Work Group met with ILCA's president Amy Spangler to learn more about their profession. Here, the LCs' lay origins became particularly problematic, for while the LC field had been moving further and further from its volunteer roots into science-based health professions (particularly nursing, but medicine and nutrition as well), the IBCLE licensure requirements did not require a specific course of study. When the Work Group learned about this, the members were "concerned" because "most physicians are unaware that certified lactation consultants have not graduated from an accredited program designed for this profession." From this point on, they worked to exert pressure on the lactation consultant profession to "improve professional standards" through selective referrals by individual physicians, by public education efforts encouraging mothers to seek out LCs with more medical training, and by publishing information on LCs for the benefit of pediatricians in the academy's newsletter.[116]

In 1997, the AAP published its updated recommendations on infant feeding, stating that mothers should breastfeed their babies exclusively for six months and thereafter up to a year or more in addition to other sources of nutrition. The recommendations garnered national media attention and prompted a widespread reaction from mothers and breastfeeding advocates. Despite efforts of the AAP, ILCA, and even LLLI (or maybe because of them), many American women viewed the increasingly adamant breastfeeding discourse warily, and in some cases with anger. In her 1997 *Washington Post* editorial, Abigail Trafford summed up some of the resistance when she wrote, "Instead of encouraging more women to breast-feed, the guideline to nurse for *at least* one year sets women up for failure." Trafford highlighted that the pressures on American mothers around breastfeeding ought to be directed, instead, at "corporate America." "So far it has done little to accommodate nursing mothers in the workplace," she argued, stating, "Breast-feeding presents logistical problems," like the need to "pump their breasts periodically and store their breast milk to be used later."[117] The identification of the breast pump alongside the increasing role of health care providers and policy makers as breastfeeding advocates became a common trope in the popular discourse by the end of the 1990s. In her 1997 multipage Sunday *New York Times* article, "Have Pump, Will Travel," Margot Slade offered an even more convincing condemnation of breastfeeding culture. Featuring the trials and tribulations of four mothers trying to combine work, life, and breastfeeding, the article began dramatically:

> Fern Drillings quit after 14 months. Jennifer Sulzberger quite after nine. Julie Goldhaum was cutting back after three weeks. Lisa Manzo stopped before she could even start. Not cigarette smoking, not diets and no, not their jobs. What these women brought to a halt was breast-feeding.[118]

From there, Slade walked readers through the awkward and exhausting routines of four working New York City moms. Each brought her own pump story. Ms. Sulzberger, a Manhattan litigation associate, recounted that she "had to lug around this electric pump with these attachments and this silly pastel-colored storage cooler with ice packs," adding that it was "another thing to clean, to keep track of and ultimately lose." And Sulzberger was perhaps the most fortunate of the mothers featured in her article because "she had an office and could avoid the 'public bathroom scenario.'" Even so, Sulzberger's tales of "trying to relax while hunched over this whirring pump" three times a day no doubt sounded less than appealing to many readers. Each story read more dramatically than the previous. Ms. Drillings, for example, a pediatric and obstetric nurse and community college instructor had different schedules every day. "On Tuesdays," she said, "I schlepped this heavy pump with me to the Manhattan VA hospital, where, during a break, I would ask total strangers if I could borrow an office, jump into what was really a closet, hook myself up to the pump, shut my eyes, think of [my son], pray that no one walked in, and pump."[119] Alongside these tales of woe was the happy and content Lisa Manzo, a secretary in a Chicago suburb who "chose not to breast-feed." "I was going back to work right away and in my job, sometimes I don't have 15 minutes to sit, let alone pump," she said. The additional benefit of Manzo's choice? The inclusion of dad in the feeding regimen. "My husband really enjoys feeding our 11-month-old, who is a healthy, happy baby, and I don't worry about pumping and storing and finding time," she said.[120]

By the close of the century, scientific studies were reporting not only that breastfeeding conveyed immunological and emotional benefits for the child, but that it made kids smarter. In 1998, *Pediatrics* published an article that stated that the *duration* of breastfeeding had a positive correlation with cognitive ability and educational achievement and that these effects were "pervasive" and "long-lived, extending throughout childhood into young adulthood."[121] Coming on the heels of the AAP's recommendations, the study received widespread media attention and the pressure on mothers to breastfeed mounted. In this context many mothers looked to the promise of breast pump technology to help free them from the growing threat of maternal failure and guilt that increasingly accompanied formula feeding. In light of this, the National Organization for Women went so far as to make a statement about the mounting pressure on mothers to breastfeed at all costs: "Some women . . . find it very difficult to breastfeed because of financial, logistical, health or other reasons. Some women are able to overcome these obstacles. Some aren't. We shouldn't use these findings to judge some mothers as good and others as bad because of their decision on this one issue of their baby's health care."[122]

The Whir of the Breast Pump

For some mothers, the breast pump never was an issue. As one New Hampshire mother said during a 1998 NPR call-in show on breastfeeding, "I breastfed for two and a half years, and it was the best thing."[123] Despite the fact that she had to pump on her work breaks "in the bathroom," she felt that pumping allowed her the freedom to continue to nurse her daughter, which she would do every day when she got home. "I never touched formula," she proudly remembered. Once her daughter began attending day care, she drove there on her lunch breaks for a midday breastfeeding. Key to her success, she added, was her local La Leche League group: "They helped me find a pump I could use at work," she said. "They helped me through all that."[124]

Other mothers, however, found the pump to be a daunting and unwelcome intrusion into their lives. A Minneapolis mother who called into the same 1998 radio show breastfed for eight months before weaning her baby. Between working full-time and her baby's increasing age, she found that her "milk supply was starting to diminish." Even after eight months she felt "guilty" because her employer offered a "lactation room and a pump that they supply," but though she tried it she found that she just "hated it." "I loved nursing," she shared, "but I didn't like pumping at all."[125] Perhaps most common, however, have been those mothers who fall somewhere in between these two extremes. In her book *The Breastfeeding Café*, Barbara Behrmann captured the experiences of mothers from across the country, many of whom had stories to share about their breast pumps. One mother, a scientist, maintained a high-profile and stressful career that required her to travel fairly often. Her husband was a stay-at-home father and fed their baby pumped milk from a bottle during the day while she worked, or sometimes for days at a time while she was out of town on business. The mother, "Dawn," spoke eloquently about her situation, discussing the realities of maintaining breastfeeding while away from one's baby, "Unlike a male colleague with a new baby who revels in a few good nights' sleep when he's away, I must get up in the middle of the night, as if my child had awakened me . . . and pump. But instead of her warm snuggly body and soft, expert lips, I'm greeted with a chair and my pump."[126] On one harrowing journey away, her infant daughter went through a growth spurt and consumed half the pumped milk she had left at home in just two days. She embarked on a frantic rush to overnight ship more milk to her anxious husband by nine a.m. the next morning. In reflecting upon her experiences, she acknowledged that she did "dream of the day when I can leave town or even leave home without my constant companion, the Medela Pump In Style." But when she addressed the question of why she pushed on regardless of these hurdles, she tellingly revealed, "Maybe it's because I sometimes feel it's my only real contribution to her day-to-day well-being."[127]

Throughout most of the back-to-the-breast movement's history, America's health care system was at best a silent witness to breastfeeding's decline and at worst an active impediment. By the 1980s, however, scientific studies of breast milk began to articulate new and culturally significant benefits for infants who were breastfed over those who received formula. At the same time, breast pump technology became more and more imbedded in mother's breastfeeding practices. Thus as science began to adamantly tout the benefits of breast milk for babies, mothers were increasingly removing themselves from the process of breastfeeding by extracting their milk. By the 1990s the advent of reasonably priced, over-the-counter electric breast pumps helped expand the role of technology in breastfeeding, a process aided by the timely emergence of the lactation consultant as a new brand of expert. Though many LCs became part of the breastfeeding movement because of an affinity with the ideology of natural motherhood, their participation in an allied health profession in the mid-1980s pulled them further and further from these roots. As the rates of breastfeeding increased and breast pumps emerged to meet the demands of working mothers with few other options, lactation consultants were poised to become not just experts of breastfeeding, but also experts of breast pumping. As health policies and recommendations urging breastfeeding took shape by the end of the century, some mothers remained skeptical of the intensifying campaign. Despite this skepticism, however, the draw of natural motherhood ideology has continued to endow breastfeeding with meaning that transcends public health arguments. Despite the strains and stresses of breast pumping, many mothers continued to do so in order to maintain a once- or twice-daily nursing session with their infant at their breast. La Leche League has remained a place that harbors mothers from across the ideological spectrum and many among the League chose to enter health fields and obtain IBCLC certification in addition to their League work. Some leaders, members, and groups became more conservative interpreters of the ideology of natural motherhood, eschewing breast pump technology as a masculine tool of oppression and relying upon conceptions of maternal instinct and mother-to-mother knowledge exchange over the more medicalized practices of lactation consultants, nurses, and physicians. The majority of breastfeeding mothers, however, worked to simply make use of the resources available that best fit the particular circumstances of their lives. Working mothers who breastfed worked out schedules with day care providers, pumped, cup fed, bottle fed, and supplemented with a variety of formulas. Mothers who had the ability and desire to stay home negotiated the demands of full-time on-demand breastfeeding, relying on pumps to unplug clogged ducts, alleviate engorgement, and give them a few hours of baby-free activity when needed or wanted. They also sometimes turned to La Leche League or other nursing support groups for moral support through the rough patches and called local LCs when even the local La Leche League leader became stumped.

By the turn of the twenty-first century, breastfeeding seemed once again an entrenched part of American motherhood even as rates of initiation remained low among younger mothers from poor economic and education backgrounds, and duration rates continued to hover far below the nation's "Healthy People 2000" guidelines. Despite these failures, few mothers by the end of the twentieth century would have discussed "formula feeding" their babies without expecting looks of disapproval or pity in return. Already by the close of the 1990s, mothers and women's rights proponents were highlighting the far-reaching structural problems that shaped women's choices when it came to breastfeeding. Pediatricians, hospitals, and nonprofits could promote breastfeeding, but when women went back to work two weeks to three months after giving birth, they faced the reality that breastfeeding might be cut short. With no real changes on the horizon regarding maternal and family policies, many breastfeeding mothers at the turn of the century looked toward a future of bottles, battles with guilt, and the all-too-often unsatisfactory whir of their breast pumps.

Epilogue

Natural Motherhood Redux

Whole Food I make that,
my baby grows big and strong,
our bond forever.
—Danielle Bridges, "Breastfeeding Haiku," kellymom.com, 27 July 2011

When I look around my daughter's second-grade class, I can't seem to pick out the unfortunate ones: "Oh, poor little Sophie, whose mother couldn't breast-feed. What dim eyes she has. What a sickly pallor. And already sprouting acne!" —Hanna Rosin, "The Case against Breast-Feeding," *Atlantic*, 1 April 2009

Breastfeeding has served as the flashpoint of the twenty-first century's "mommy wars."[1] Over half a century since its slow return first began as a push back against a scientific ideology and a wave of antimaternalism, breastfeeding is now widely understood as a divisive issue not between mothers and the health care system, nor parents and the employment sector, but between mothers themselves. In her 2009 piece in the *Atlantic*, writer Hanna Rosin mounted a "case against breastfeeding," arguing that "in certain overachieving circles, breast-feeding is no longer a choice—it's a no-exceptions requirement, the ultimate badge of responsible parenting."[2] In her exposé Rosin offered a challenge to the medical literature's ambiguity on the question, Is breast best? and even called out the breast pump as anathema to the message of breastfeeding as natural, despite the vast majority of breastfeeding moms who have come to rely on the device. "When people say that breast-feeding is 'free,' I want to hit them with a two-by-four," she wrote emphatically, adding, "It's only free if a woman's time is worth nothing."[3] Rosin's piece splashed in an already roiling public discourse on breastfeeding. The discussion that followed the *Atlantic*'s publication of the piece fell into predictable and familiar lines of debate. Most women who responded expressed a desire to feel that breastfeeding was an option, one they could embrace or reject depending on the individual circumstances of their lives, including careers and the level of difficulty they encountered in trying to breastfeed.[4] Still, there remained those who held steadfastly to the sentimental and moral arguments for a "natural" breastfeeding experience, a decidedly sentimental vision of nursing epitomized in the haiku at the start of this epilogue.[5] Despite the intelligent and well-researched arguments of people like Rosin, however, the debate over breastfeeding continues to remain inadequately framed and inherently divisive in ways that are both frustrating and harmful. Breastfeeding is not now, nor has it ever been, merely a matter of personal choice and in continuing to cast it as such, from either a feminist or maternalist standpoint, leads only to bitter public rivalries and ineffective policy making.

Where Are We Now?

Since the 1990s, rates of breastfeeding in the United States have continued to climb alongside public health efforts aimed at increasing the number of women who choose breastfeeding. This upward climb has no doubt been spurred on by the efforts of the Healthy People initiatives set forth by the Surgeon General's Office since 1979, which have included breastfeeding as a preventative health measure.[6] As is the case with any statistic, however, reported breastfeeding rates can obscure a larger reality. The Centers for Disease Control and Prevention (CDC), which reports on breastfeeding data collected from several different surveys, has shown consistent gains in the numbers of mothers initiating "any breastfeeding" since 2000.[7] As of 2010, for example, the CDC reported that the Healthy People goal for breastfeeding initiation had been met, climbing up six percentage points to 75 percent.[8] Exclusive breastfeeding for any duration, however, has made few gains. As of 2010, the CDC could only report that 13 percent of breastfeeding mothers were doing so exclusively for six months.[9] Other data indicates a similar lack of success in increasing the rates and duration of exclusive breastfeeding. Between 2003 and 2010, for example, the percentage of breastfed children who received formula supplements in the first two days of life fluctuated between 22 percent and 26 percent, while the percentage who received supplements during the first six months of life dropped only slightly, from 47 percent to 43 percent.[10] Despite repeated statements by the American Academy of Pediatrics since 1997 on the importance of breastfeeding exclusivity, reports continue to demonstrate that the percentage of mothers who are both willing and able to follow this advice has hit a wall. Even more troubling for breastfeeding advocates are the continuing disparities in breastfeeding rates and experiences along racial lines. In 2008, a rate of initiation of breastfeeding among black mothers was 58.9 percent, significantly lower than that among white (75.2 percent) and Hispanic (80.0 percent) mothers.[11] Continually puzzled over why mothers would choose to deprive their children, themselves, and society of the "significant benefits of breastfeeding," researchers, with a collective scratch of their heads, have attempted to explain the nature of the limits to breastfeeding's expansion by focusing on various forms of the question, Why do mothers stop [or never start] breastfeeding?[12] Such inquiries have yielded a host of familiar answers, including everything from sore nipples to "interference with mothers' lifestyle."[13] Despite repeated findings that the women most likely to stop breastfeeding within the first month of their infant's life tend to be those who are "younger" and who have "limited socioeconomic resources," few successful efforts have been made to cast breastfeeding in any light other than as a personal health choice.[14]

There are other issues beneath the surface of these recent data, as well, particularly the matter of how researchers define their subject. NIS questions about "exclusive breastfeeding" were not adequately refined until after 31 December 2005, when the survey began utilizing the question, "How old was [child's name] when (he/she) was first fed formula?"[15] As of 2013, however, the NIS does not distinguish between breastfeeding and being fed breast milk; research today continues to largely ignore the actual process by which an infant is fed. Only recently have scientists and clinicians begun to question the conflation of breast milk feeding and feeding at the breast. In 2010, Sheela R. Geraghty, from the Cincinnati Children's Hospital's Center for Breastfeeding Medicine, and Kathleen Rasmussen, from Cornell University's Division of Nutritional Sciences, brought this issue to light in a letter to the editor sent to the journal *Breastfeeding Medicine*. They thoughtfully asked, "Is a child fed at the mother's breast and also fed bottled pumped milk by another caregiver 'exclusively' breastfed?"[16] Their question suggested that for research purposes it seemed "important to distinguish between feeding directly at the breast and feeding pumped milk because these behaviors may have different effects on the health of mothers and children."[17] Despite these discussions, the American Academy of Pediatrics Section on Breastfeeding's 2012 policy statement makes no mention of distinctions between breastfeeding and breast milk feeding.[18]

When exclusivity, extraction, and duration are taken into account, then, it becomes clear that the health care community's desire to see progress in breastfeeding over the past several decades has contributed to an obfuscation of some fundamental and underlying issues. Current guidelines aside, efforts to increase breastfeeding in the last several decades have contributed to the rise of a culture of "any breastfeeding." The rhetoric of individual choice that continues to influence breastfeeding education, support, and outreach, furthermore, dilutes the complexity and depth of the problem of mothers' ongoing struggles to breastfeed. From the perspective of mothers, of course, opinions and experiences vary widely, though in the public discussion on breastfeeding one can find an evolving consensus that breast pumping and breastfeeding are two different and unique processes, though they are often used in combination.[19] Another thread in recent breastfeeding discussions reveals, too, that lactation consultants, who have lacked cohesion arguably since their profession first formed, continue to bring wildly different beliefs and perspectives to their encounters with mothers. Just as mothers in the postwar years might have traded horror stories about insensitive physicians and nurses, many mothers today can now swap stories about negative experiences with lactation consultants, even as many others have been greatly helped by an LC.[20] And despite the peaceful vision of breastfeeding championed under the ideology of natural motherhood, stories written by and about mothers have filled newspaper columns as well as the pages of magazines, books, editorials, and blogs, attesting to the trials and tribulations that have accompanied the "choice" to breastfeed for the past four decades.

The time is ripe for scholars, mothers, health care professionals and anyone who cares about the status of women in America to confront the issues underlying the rise of modern breastfeeding evangelism in the medical community, the recent backlash from some mothers who have articulated a resistance to it, and the ambivalence of policy advisers and makers alike who have focused on stunted efforts to encourage employers to accommodate breast pumping, while doing very little to support breastfeeding mothers at the national level. In the broader world of reproductive justice, women of color have led the move away from a language of individual "choice" more generally in favor of a more honest and inclusive social perspective. Organizations like the SisterSong Women of Color Reproductive Health Collective and Asian Communities for Reproductive Justice have argued that "the focus on and orientation towards individual rights and individual responsibility, as they relate to articulation of reproductive health and women's choice, reinforce the broader systems of political, economic, and cultural hegemony that privilege and maintain racial [and I would argue, economic] stratification in the United States."[21]

Much as these activists and scholars have called for a shift away from a language of choice in reproductive justice more broadly, the place of breastfeeding in these discussions often remains on the sidelines amidst the more life-and-death issues of abortion, pregnancy, and birth control.[22] Debates about breastfeeding, however, can only benefit from engagement with a reproductive justice perspective arguing that women ought to "have the economic, social, and political power and resources to make healthy decisions" for themselves, their families, and their communities.[23] As a handful of authors have been increasingly willing to point out, there are far deeper issues behind the stagnating breastfeeding movement, issues that are fundamentally about how we as a nation define and value motherhood and women.[24]

Moving Forward

The history, analysis, and conclusions laid out in this book aim to provoke conversations and policy revisions regarding infant feeding and motherhood. The splits within the breastfeeding and women's health communities in the 1970s and 1980s had a polarizing effect on ideologies of "the natural," and arguments in the rhetoric of natural motherhood became increasingly affiliated with a conservative effort to resist the expansion of women's biological, economic, and political rights. Global scrutiny on breastfeeding and children's health by the 1990s helped shine a spotlight on the situation in the United States and by the end of the twentieth century, medical and public health organizations had taken up the cause to increase breastfeeding across all demographics. In doing so, however, breastfeeding support efforts in the last two decades have not approached breastfeeding as an embedded, embodied, and cultural aspect of motherhood, but have instead followed the path laid out by the debates of earlier decades over working motherhood and personal choice. Campaigns to increase breastfeeding have since relied upon a strategy of providing scientific facts about the discrete health benefits for the breastfed child, and to some extent for the mother as well, and providing technological assistance (such as breast pumps and lactation consultants) to help overcome immediate obstacles. Much like physicians' arguments that focused on the gastrointestinal health of breastfed babies in the 1940s, however, this kind of support has consistently done little to modify the structural constraints impacting women's experiences with breastfeeding. In fact, public health efforts over the past several years to expand breast milk feeding have sparked a growing backlash by mothers and public health analysts, who are often justifiably skeptical of expert claims of breast milk's vast superiority over formula.[25]

This is not the first analysis to point out these discrepancies in contemporary breastfeeding discourse. Bernice Hausman and Penny Van Esterik have argued that "social change for women can occur on the basis of biological discourses about breastfeeding, as long as they are articulated within feminist frameworks." Furthermore, Hausman has suggested that the reason this feminist articulation of breastfeeding has faltered is that the "biological perspectives on breastfeeding have been produced in the context of scientific motherhood," and therefore have produced "discourses about breastfeeding as an embodied maternal health benefit . . . in the context of the physician's authority to regulate infant feeding."[26] These perspectives, however, have become increasingly silenced in the most recent spate of writing and scholarship on breastfeeding, which has cast breastfeeding as part of a patriarchal hegemonic medical discourse aimed at maintaining a gender and sex hierarchy.[27] In arguing that the back-to-the-breast movement originated not within the framework of scientific motherhood, but within an alternative and woman-centric ideology of the natural, however, I suggest that there is no inherent reason why biological arguments for maternal and women's rights should be slated as conservative, passive, or accommodationist. Rather, as female experts and mothers alike have demonstrated throughout the past century, arguments built on a concept of embodied maternal knowledge can and have been mobilized for the empowerment of mothers. That this long and slow movement has not widely influenced social or public health policy making up to this point does not suggest that it cannot do so in the future.

The Contested Breast Pump

Feeding a baby exclusively at the breast is a distinct process from feeding a baby using complementary or supplementary formula and/or expressed breast milk in a bottle. This is not to denigrate the multitude of processes both now and in the past that mothers have relied upon to feed their babies in a manner that works for them. But it is a notable distinction to make that pumping milk and feeding it to a baby and feeding a baby at the breast are two very different processes that, in turn, yield a set of very different experiences and meanings for those involved. Breastfeeding itself cannot be taken for granted as a singular process; any mother who has nursed more than one baby will likely recount that each one was different. Furthermore, any mother who nurses for any length of time will also no doubt confirm that the process of breastfeeding a single baby changes dramatically over time. The processes involved in infant feeding, like any human relationship, are dynamic and cannot be reduced to a set of standardized instructions. In the end, the implications and consequences of the particular process one uses to feed a baby are difficult to distill to any generalizable set of recommendations or guidelines. Advising mothers to feed their babies a certain way for a certain length of time ignores the complexity and importance of the process itself and it results in dissatisfied mothers and widespread maternal guilt.[28]

What is true is that if a mother breastfeeds, she encounters a set of options that could, in the aggregate, contribute to a reinterpretation of motherhood in American society. A mother who can feed her baby entirely from her own body, for example, has the ability to reduce her participation in a capitalist consumer-based culture by limiting her reliance on commercial formulas and other infant feeding technologies. Mothers who can breastfeed also have the option to claim knowledge of their infants' care and nutrition for their own, removing it partially from the direct oversight of the medical system. Much as La Leche League first helped articulate in the 1950s, a mother who breastfeeds today can still feel a sense of embodied power and purpose, a meaning to her biological motherhood that transcends the simple exchange of nutrients. This is not to say that breastfeeding automatically and magically bestows these things, but it does give rise to a unique set of possibilities that are worth our consideration.

Similarly, there is nothing inherently feminist or antifeminist about breast pump technology. This modernized tool has, in its fairly brief history, helped to integrate breast-milk-feeding mothers more fully into the employment sector without fundamentally restructuring the workplace or our definitions of productivity, labor, motherhood, or value, and it has fallen in line with a culture of consumer technologies that promise less work for women, but nearly always result in the opposite.[29] Through this lens, breast pumps and their alignment with a rhetoric of "choice" have obscured the degree to which women continue to operate within a gender-biased system that simultaneously exploits their labor as workers and as mothers.[30] This is not a critique of mothers who breast pump, nor of breast pump technology. As I have also argued, breast pump technology can be and is often experienced as a tool of individual empowerment that allows mothers to exert agency and meet the multifaceted challenges of their lives. Additionally, even the most persistent breastfeeding mother can and often does benefit from the use of a device that compensates for an infant on a nursing strike, or who wants to (or needs to) be away from her infant for more than a few hours at a time. The pump, however, like most technologies, can serve more than one master, and context matters. In America, the pump has facilitated conversations about breastfeeding and working motherhood that focus on individual circumstance and choice rather than on any overarching issues about maternal rights. In other countries, where cultural and political contexts have differed, pump technology exists alongside national policies that guarantee paid maternal (and in some cases paternal) leave for up to a year so that mothers need not make the choice between breastfeeding and breast pumping or formula feeding with the threat of unemployment, loss of income, stress, and loss of stature that often accompany the choices of mothers in this country.[31] When framed as a matter of individual choice, breast pumping will likely continue to stand in for breastfeeding. And, as more and more critics have highlighted, once you put it in a bottle, the differences between feeding a baby formula or mother's milk diminish, especially from the mother's perspective of process and labor.

Lactation versus Breastfeeding

The history of breastfeeding's return suggests that there is an important distinction to be made between lactation support and breastfeeding support. Lactation suggests a purely physiological process: a breast lactates, but a mother breastfeeds; one is a passive construction, the other active. The physicians who advocated "breast is best" in the mid-twentieth century learned this distinction too late, as they watched breastfeeding fall further and further out of favor despite their efforts to manage it medically. Without access to the female-support networks offered by groups like La Leche League or the Boston Association for Childbirth Education, women who set out to breastfeed throughout the postwar years were far more likely to end up formula feeding than not. Advice books and approving doctors could only get a woman so far when the broader context in which she lived supported bottle-feeding. Despite the remarkable shift in favor of breastfeeding over the last fifty years, however, mothers today experience a far more medicalized form of lactation support than they did even three decades ago. Every mother and every baby are, in fact, different. The effort to standardize lactation support and embed it within traditional medical institutions may have helped increase the number of babies who receive some breast milk, but it has not addressed the individuality of the nursing pair or the embodied and socially meaningful experience of the mother. Some mothers and babies require little outside assistance; others do everything "right" only to find that they have issues like an overactive let-down reflex; still others are simply discouraged when the now standard three- to five-day post-hospital discharge check-in reveals that their baby hasn't regained his or her birth weight.[32] Wrapped in a postpartum, sleep-deprived, hormonal haze, mothers confronted with any challenge to their ability to feed their babies are wont to fall into a tailspin of self-doubt. The wariness with which postwar breastfeeding supporters approached medical intervention in breastfeeding reflected their experiences with these same issues almost a century ago. As a result, the early back-to-the-breast movement formed to resolve women's breastfeeding problems without having to rely on a medical system, which they knew would fundamentally work against their efforts. Mothers in difficult situations obtained donated milk from their support groups, rented hospital-grade breast pumps, and most importantly received consistent encouragement and friendship throughout the ups and downs of breastfeeding and motherhood.

Although many mothers today have the option of meeting with a lactation consultant, these encounters now occur largely within the very same structures that historically thwarted women's efforts to breastfeed. In expanding hospital-based lactation support to all women, we ought to consider, then, what may have been left behind. Much as in the history of the medicalization of childbirth, mothers today have traded authority and social support for access to the specialized knowledge and technologies of modern lactation expertise. Encounters with medical experts, as history has shown, however, do not always (or even usually) provide the supports traditionally offered by female institutions. Much as mothers used to look to their nurses for female support with breastfeeding, they now look to lactation consultants. All too often, however, differences in perspective, experience, background, and kinds of knowledge plague these modern-day interactions between patients and health care workers just as they did in the past. In imagining what a more effective system of breastfeeding support might look like, then, we ought to consider the continuing desire by mothers to have access to ongoing social support networks that exist outside of medical institutions.

Our Toxins, Ourselves

In 1977 the Environmental Defense Fund, seeking to build off of the groundswell of concern and anxiety surrounding breast milk contamination, published a report provocatively titled *Birthright Denied: The Risks and Benefits of Breastfeeding.* Throughout its fifty-five pages, the author, EDF research associate Stephanie Harris, reported alarming concentrations of chemicals that she characterized as posing a "highly significant" cancer risk. During Senate subcommittee hearings that same year on the contamination of mother's milk, Harris put her findings in even starker terms, citing data that showed that women who do not breastfeed had higher rates of breast cancer than those who did—"We are talking here about the tradeoff between the risk of cancer to the child or risk of cancer to the mother," she said. "This is a terrible decision for any woman to have to make."[33]

In the late 1970s, mothers watched this debate unfold in the wake of the passage of the Toxic Substances Control Act, nearly five years after DDT use had been largely banned, and in the post–Clean Water and Clean Air Act era. So many gains had been made and yet still their environment and their bodies remained marked by these chemicals and by the uncertain risks they posed. Rife with concern, environmentalists and maternal activists alike slowly shifted away from organized action to a discourse of personal choice and individual risk. The 1977 Senate subcommittee hearings presided over by Senator Edward Kennedy reveal a shift away from a belief in a structural and legislative approach to remediation at the community and national levels to a modern public health language embedded in an ideology of scientific motherhood. Armed with enough scientific knowledge about what was in their breast milk and the risks and uncertainties it posed, mothers, it appeared, would be left to decide for themselves between the risks of breastfeeding or not breastfeeding. It was a choice weighted with women's fraught history as marginalized bearers of the burdens of a patriarchal and technological modernity. With bodies and breast milk acting as reservoirs of society's waste products, mothers were asked if they would choose bodily quarantine or face passing on their burden to the next generation.

During her opening statement at the 1977 subcommittee hearings, EDF's Stephanie Harris helped outline the individual risk framework that would come to define the issue of contaminated breast milk for at least the next twenty-five years, "The question which we are here to discuss today is whether or not it is safe to breast feed." "I was confronted with this in a very personal way 15 months ago when I gave birth to my first child," she stated. "At that time, I realized that I did not have adequate information either on the benefits or the risks of breast feeding to answer this question, and yet with the information which I did have available to me I chose to breast feed my baby. I certainly do not regret it."[34] Although she had wished for better data, Harris nonetheless analyzed the risks she knew of and decided, for herself and for her baby, that the benefits of breastfeeding outweighed the risks. Later on in her testimony, Harris revealed an even greater belief in personal choice and agency in this issue when asked by Senator Kennedy how she calculated the risk to her baby: "The benefits for me probably outweighed the risks, primarily for one reason, and that was I had been on a low-fat diet for a number of years in order to try to minimize these contaminants in my breast milk. . . . Some scientists say this diet should be followed even from the onset of puberty if a woman wants to purify her system enough so that she will be able to nurse her baby."[35] Although Harris went on to suggest that the FDA and EPA "eliminate these persistent chemicals from the food and water supply," she also recommended that women eliminate the consumption of meat, fish, and high-fat dairy products from their diets and suggested that breast milk toxin screenings should become a standard part of postpartum care. Under the EDF's proposal, all mothers were to be offered the option of a breast milk screening. "Armed with the results of the analysis," they wrote, "the woman can turn to our booklet as a basis for calculating the risks of breastfeeding against the benefits." When pressed as to whether she would recommend against breast feeding in some cases, Harris suggested that in the case of high toxin levels, mothers could again change their diets and follow a minimized breastfeeding routine—one in which mothers might breastfeed only once or twice a day in an effort to obtain the immunological benefits of breastfeeding while decreasing the risk. Later in his own testimony, Dr. Robert Harris, Stephanie's husband and the associate director of EDF's Toxic Chemicals Program, would state their position even more bluntly: "Whether or not the benefits outweigh the risks," he said, "it is obviously a highly individualized decision."

Despite an obvious interest in ongoing federal efforts to clean up the environment, the EDF revealed in its testimony both a growing pessimism about the relationship between environmentalism and breastfeeding activism and a rhetoric of "every woman for herself." While the language of "knowledge is power" fit into broader discourse in this decade, when it came to breast milk toxins the idea that mothers should be responsible for both knowing about their toxic breast milk and either minimizing their exposure in the first place or remediating their bodies through personal consumption choices placed an inordinate amount of unremunerated labor along familiar gendered lines. The implications of this shift would give rise to a growing subculture of natural motherhood, an ideology informed by ecological ways of knowing, an embrace of maternity as a salient but apolitical category of identity, and a belief in the power and obligation of the female self to construct unspoiled pockets of nature within the bounds of the body, the family, and the home. At the same time, the negative consequences of the rise of this discourse would also travel along traditional boundaries of race and class, burdening those with the least access and the fewest resources with the greatest environmental catastrophes and a minimal ability to alter personal circumstance. One of the more dramatic illustrations of this would come in the mid-1990s when, after being warned by health officials in Canada and the United States about widespread contamination of arctic fish, whales, walruses, and seals, many Inuit mothers would abandon breastfeeding and turn instead to a formula of Coffee-mate mixed with water.[36]

In the late 1990s, concerns about environmental contamination and bodily toxins continued to percolate throughout the breastfeeding community. Concerns over the contamination of fish, for example, lingered in discussions on LACTNET. Dedication to the superiority of breast milk characterized these discussions, as they had in earlier decades, with lactation consultants and La Leche League leaders articulating a common refrain: "We know breastfeeding is best . . . but we need to clean up the planet too."[37] Within small pockets of the breastfeeding support community a rhetoric of shared responsibility and universal burden would continue to persist, but with few long-term organizational successes.

The maternal peace and environmental activism efforts of the 1960s culminated in the 1970 celebration of the first Earth Day. As part of the events taking place throughout the nation on 22 April that year, mothers coordinated a grassroots effort to bring the topic of breast milk contamination into the spotlight. Environmental contamination served, for a time, as a uniting issue, bringing together breastfeeding mothers from across a diverse spectrum of religious and political views. By the end of the 1970s, however, the discourse surrounding infant feeding had become thoroughly infused with the rhetoric of personal and consumer choice and bodily autonomy.

The move away from community- and organization-level activism around breast milk contamination, however, has helped to shift the burden of blame and remediation from the producers and regulators of environmental toxins onto individual women's bodies. Much like the American public health care system more broadly, modern breastfeeding advocacy and support focus on individual choice, action, and personal risk, rather than addressing the widespread and fundamental environmental conditions and structural obstacles that limit the success of breastfeeding and diminish the quality of breast milk. In the specific case of bodily toxins, the unwillingness to embrace the connections between bodies and their environments at the level of public health policy has not only undermined the ability to effect the regulatory and policy-level changes but also relegated the potentially radical politics of maternal health to the private sphere and inscribed gendered, racial, and class inequalities deeply onto the bodies of women and children.

Despite this rather dire story, there have been glimmers of a faint but burgeoning twenty-first-century maternalist and environmentalist alliance. Visible in the 2005 founding of the Bay Area group Moms Organized to Make Milk Safe (MOMS), a new generation of mothers has become increasingly politicized by recent crises surrounding toxic chemicals in their breast milk—this time found hiding not in our foods, but in our furniture. One day, Bay Area mother Mary Brune sat down on her couch to nurse her daughter, which is how she first learned via a news broadcast about the presence of flame retardants in sofas and their habit of winding up in women's milk. "I was scared and angry and motivated," she recalled. In response, she and three of her friends who were also moms and who had various ties to the environmental movement decided to found MOMS, a group dedicated to educating the public and lobbying for legislative change on the issue.[38] MOMS gained national visibility with Brune's appearance as a *Huffington Post* blogger in 2007, and after the group launched a campaign lobbying for the ban of the chemicals in California, the state capitulated.[39] It remains to be seen if MOMS will continue to be an anomaly, or if it suggests the possibility of a renewed backlash against the personal choice framework that has grown to dominate breastfeeding discourse since the late 1970s.

Natural Motherhood Redux

When the back-to-the-breast movement began in the mid-twentieth century, it embraced an ideological belief in natural motherhood, a construct of maternity that relied upon a kind of moral biology built on the sciences of instinct, evolution, psychology, and animal and human behavior. This ideological framework not only kept breastfeeding knowledge alive during a period of decline; it fostered a wide network of experts, mothers, and supporters alike who created an alternative model of breastfeeding science, one built on women's bodily knowledge and the scientific study of a body enlivened by the mind. When this midcentury movement confronted second-wave feminism and an organized environmental movement, the ideological orientation became increasingly associated with a conservative position that stressed women's place in the domestic sphere. In the wake of this division, it became increasingly implausible for progressive and liberal-minded supporters to make arguments about motherhood from a biological or natural perspective. In the public discourse, biological essentialism became a prison, and "natural" processes no longer offered a clear route to maternal empowerment.

The breast pump in this context offered one possible solution to many of the problems facing breastfeeding and working mothers. By the end of the century, connections between breastfeeding and natural motherhood had loosened as breast pumps helped to increase the time and distance separating a mother and nursing baby and allowed the creation of a new model of breastfeeding. Examining this history, however, suggests that rather than witnessing the resolution of the tensions between natural motherhood and modern breastfeeding, mothers who breastfed at the end of the twentieth century continued to experience structural obstacles, physiological difficulties, and emotional battles reminiscent of their mid-twentieth-century forebears. The history of breastfeeding's return highlights the fact that late twentieth- and early twenty-first-century efforts to institutionalize breastfeeding have lacked an acknowledgment of the movement's roots as a woman-centric effort built on nonmedicalized models of maternal instinct, authority, and emotion, as well as a belief in the inherent value of "the natural." Recent arguments against breastfeeding moralism and breast pump technology highlight the extent to which the practice has, in recent years, come to be seen by a growing number of women as part of a medical and public health effort that ignores the needs and wants of mothers themselves. Despite this, there continue to be women who advocate for breastfeeding as part of a way of living that embraces bodily rhythms, instinctual responses, and "the natural." While it seems likely that mothers will continue to find personal value in their efforts to breastfeed for many years to come, it also seems not unlikely that the recent pushback against breastfeeding advocacy could lead us to yet another reversal in American infant feeding trends. The goal of the analysis provided here is to help us to think more expansively about the role of breastfeeding in our society in the past, present, and future. Ultimately it is my hope that whatever the coming generations decide to make of breastfeeding, they do not proceed in ignorance of what it has meant in the lives of the many women who have struggled to bring it back.

Acknowledgments

This book is the result of years of intellectual, logistical, and emotional challenges matched in scope only by the support and assistance of the network of friends, family, colleagues, mentors, and institutions that have helped me to meet them.

I have first and foremost to thank graduate students, staff, and faculty in the History and Sociology of Science Department at the University of Pennsylvania for providing a rigorous, challenging, and supportive environment in which this project first took shape. I am especially indebted to David Barnes, who encouraged me to pursue this project when I first broached it as a topic. Susan Lindee pointed me in the right direction, helped me to envision what a project like this might look like, and gave me valuable feedback throughout the early stages of the book. Beth Linker has read and critiqued countless versions of this work over the years, and her persistent interest and support in this project has helped carry me through more than a few rough patches. The mentorship of Kathy Peiss in the History Department at the University of Pennsylvania was indispensable early on in helping me identify the central questions that now underpin much of my work. Opportunities to present pieces of this project to Patricia D'Antonio, Julie Fairman, Cynthia Connolly, Jean C. Whelan, and others at workshops held at the University of Pennsylvania's Barbara Bates Center for the Study of the History of Nursing helped me think more expansively about the important place of nursing history in my work.

Thoughtful feedback from audiences, panelists, and commenters at the annual conferences of the American Association for the History of Medicine, the American History Association, the American Society for Environmental History, the History of Science Society, and the Society for the History of Technology helped me work through tough questions, pointed me toward important resources, and put me in touch with a broad network of scholars doing interesting and relevant work. I am particularly indebted to Judith Walzer Leavitt for nudging me down the right path more than once and to Wendy Kline for helping me to think about the connections between the history of childbirth and breastfeeding. Rebecca Jo Plant read excerpts and provided invaluable comments on drafts of articles and conference papers; her intellectual interest and professional mentorship throughout this process has been more important than she likely knows.

There are more people who have helped shape my scholarship than I can reasonably mention in this brief acknowledgments section, but I would be woefully remiss if I did not thank those who provided direct support in the form of source materials and advice throughout the life of this manuscript. Ellen Stroud, at Bryn Mawr College, encouraged me to think about the connections between the history of breastfeeding and environmental history in ways that turned out to be critical to how I conceptualized this book. Kristoffer Whitney read excerpts, provided me with historical treasures from his grandparents' attic, helped me locate hard-to-find articles and images in the collections of the University of Wisconsin, Madison, and most importantly offered moral support, comic relief, and friendship throughout the process of writing this book. Eric Hintz at the Smithsonian's Lemelson Center for the Study of Invention and Innovation provided much-needed advice on how to conduct my research into the history of breast pump technologies. Samantha Muka and Dominique Tobbell have provided an endless source of moral support and critical feedback on this work through their generous offers to read excerpts, attend conference talks, and commiserate on the ups and downs of academic life. Meghan Crnic read early drafts and provided me with helpful reading suggestions along the way.

My colleagues in the Department of History at Mississippi State University have helped to provide a positive and supportive environment in which to complete the research and writing of this book. In addition to funding important archival excursions, the department at MSU allowed me valuable research leave which helped make the timely completion of my manuscript possible. I would like to particularly thank my colleague Julia Osman, who read drafts of conference papers, articles, and countless excerpts from my manuscript and held me accountable for my self-imposed deadlines. Throughout the course of our weekly writing sessions, I gained far more insight into the history of France's military endeavors than I ever thought possible, while she graciously learned more about the history of breastfeeding than she likely ever cared to.

The research for this book required the helpful and very necessary assistance of many archivists and librarians who granted me access, located hidden treasures, and answered countless questions. Though too many to list individually by name, I am deeply indebted to professionals at the Schlesinger Library at the Radcliffe Institute for Advanced Study, DePaul University Special Collections, Syracuse University Special Collections, the American Academy of Pediatrics, the Library of Congress, the University of Pennsylvania Archives, the Pediatric History Center Collection at the American Academy of Pediatrics, Columbia University Archives, and the University of Oregon Special Collections and Archives. I would not have been able to unearth the history of the modern breast pump without the kind and selfless assistance provided by friends and family of Einar Egnell. I am particularly indebted to Rolf Egnell and Olle Levan who graciously recounted stories, transcribed and typed letters, and exchanged e-mail and postcards with me throughout my research process. Assistance in the form of references and contact information provided by representatives at both the Medela and Ameda corporations was also essential to completing my work on the breast pump. Additionally, without the words and stories of mothers this book would be far less interesting, so I am especially indebted to the mothers who shared their time and experiences with me in interviews. While only a small handful of interviews made it into the final manuscript, the countless conversations I've had with mothers since I began this project have greatly influenced my perspective and the questions I've sought to answer.

I am very grateful to Karen Merikangas Darling, my editor at the University of Chicago Press, who has guided me through the publication process with patience and encouragement. Her calm responses to my sometimes panicked e-mails always helped me put things in perspective and allowed me to focus on the tasks at hand, while Evan White, her editorial associate, helped answer my many, many questions about things like permissions and fair use.

I never fully understood why authors left family until the very end of their acknowledgments, but now that I am writing them myself it makes quite a bit of sense to save the biggest thanks until last. My large and vibrant extended family has always played a significant role in my life, and throughout the writing of this book their support and love has been a source of strength and inspiration. I want, especially, to thank my grandmother, Barbara Crandley Hudak, whose interest in family history and storytelling ended up in the pages of this very book. My family, and particularly my parents, Mike and Carol Martucci, always encouraged me in my dedication to education and academia, even when I made the dubious choice to leave my very respectable BA in biology behind in favor of a career in the humanities. My brother and sister have patiently put up with an older sister who has spent much of the past decade being quite boring while buried under piles of books and articles.

Aside from me, it is well within reason to say that no one has pored over every page and every sentence of this manuscript as many times as my dearest friend and wife, Abby, who has surely by now earned the rights to an honorary PhD in history. Her patient willingness to share our lives with this project for the past six (or more) years is something for which I owe her more gratitude than I can ever adequately express. Finally, I am thankful to my daughter, who has not been here very long yet, but who has already taught me more about motherhood and love than I could ever learn from books.

Notes

Introduction

1. Martha Pugacz to Hazel Corbin, 22 September 1956, Members Files—Martha Pugacz, La Leche League International Records, Special Collections and Archives, DePaul University Library, Chicago (hereafter LLLIR).

2. See, for example, Rima D. Apple, *Mothers and Medicine: A Social History of Infant Feeding, 1890–1950* (Madison: University of Wisconsin Press, 1987); Adrienne W. Berney, "Reforming the Maternal Breast: Infant Feeding and American Culture, 1870–1920" (PhD diss., University of Delaware, 1998); Janet Lynne Golden, *A Social History of Wet Nursing in America: From Breast to Bottle* (New York: Cambridge University Press, 1996); Bernice L. Hausman, *Mother's Milk: Breastfeeding Controversies in American Culture* (New York: Routledge, 2003); Sally McMillen, "Mothers' Sacred Duty: Breast-Feeding Patterns among Middle- and Upper-Class Women in the Antebellum South," *Journal of Southern History* 51, no. 3 (August 1985): 333–56; Marylynn Salmon, "The Cultural Significance of Breastfeeding and Infant Care in Early Modern America," *Journal of Social History* 28, no. 2 (1994): 247–69; Kara W. Swanson, "Human Milk as Technology and Technologies of Human Milk: Medical Imaginings in the Early Twentieth-Century United States," *Women's Studies Quarterly* 37, nos. 1–2 (2009): 20–27; Paula A. Treckel, "Breastfeeding and Maternal Sexuality in Colonial America," *Journal of Interdisciplinary History* 20, no. 1 (1989): 25–51; and Jacqueline H. Wolf, *Don't Kill Your Baby: Public Health and the Decline of Breastfeeding in the Nineteenth and Twentieth Centuries* (Columbus: Ohio State University Press, 2001).

3. Apple, *Mothers and Medicine*, 3–19, 150–68.

4. Ibid., 152–55; Jacqueline H. Wolf, "Low Breastfeeding Rates and Public Health in the United States," *American Journal of Public Health* 93, no. 12 (December 2003): 2000–2010, 2004.

5. Christina Bobel has written about "natural mothers" as a coherent group,

identifiable in contemporary twenty-first-century American life. She argues that "today's natural mothers are . . . a contemporary variant of the female moral reformers" of the nineteenth and early twentieth century. As such, she argues that they "embrace a similar politics of accommodation," in which they "resist certain capitalist structures and attempt to wrest control from institutions and challenge 'experts' they perceive as threatening . . . [but] do not challenge the structure and content of gender relations." My argument departs from this analysis first and foremost by suggesting natural motherhood has functioned more broadly as an ideological framework, just as scientific motherhood has. It has and continues to influence the construction of knowledge and the behaviors, ideas, and experiences of mothers, policy makers, and health care providers whether they choose to embrace the ideology or not. Furthermore, Bobel casts natural motherhood as part of a spirit of "accommodationism," but I argue that there was nothing inherently antifeminist nor accommodationist about natural motherhood ideology. The fact that it has become associated with a conservative set of arguments about women's reproductive and domestic worth is the result of historical circumstances, which are the subject of this book. See Christina Bobel, *The Paradox of Natural Mothering* (Philadelphia: Temple University Press, 2002), 46–47.

6. Elaine Tyler May, *Homeward Bound: American Families in the Cold War Era* (New York: Basic Books, 1988), 136.

7. On the term "mother-guilt," see Diane Eyer, *Mother-Guilt: How Our Culture Blames Mothers for What's Wrong with Society* (New York: Random House, 1997). For a larger analysis on the blaming of mothers for social ills, see Molly Ladd-Taylor and Lauri Umansky, eds., *Bad Mothers: The Politics of Blame in Twentieth-Century America* (New York: New York University Press, 1998); and Nancy Pottishman Weiss, "Mother, the Invention of Necessity: Dr. Benjamin Spock's Baby and Child Care," *American Quarterly* 29, no. 5 (Winter 1977): 519–46, 520.

8. Alice Kessler-Harris has observed that between 1970 and 2000, "the proportion of wage-earning mothers of children under six doubled . . . rising from 30 to more than 63 percent." See Alice Kessler-Harris, *Out to Work: A History of Wage-Earning Women in the United States*, 20th anniversary ed. (Oxford: Oxford University Press, 2003), 325–26.

9. Niles Rumely Newton, *Maternal Emotions: A Study of Women's Feelings toward Menstruation, Pregnancy, Childbirth, Breast Feeding, Infant Care, and Other Aspects of Their Femininity* (New York: Paul B. Hoeber, 1955).

10. Apple, *Mothers and Medicine*, 35–36; Wolf, *Don't Kill Your Baby*, 187–88.

11. On the culture of medical expertise and technologies as masculine, see Joan Cassell, *The Woman in the Surgeon's Body* (Cambridge, MA: Harvard University Press, 2000), 100–128; Regina M. Morantz-Sanchez, *Sympathy and Science: Women Physicians in American Medicine* (New York: Oxford Univer-

sity Press, 1985); and Ruth Oldenziel, *Making Technology Masculine* (Amsterdam: University of Amsterdam Press, 2004).

12. On infant feeding, see Apple, *Mothers and Medicine*, 3–19; Wolf, *Don't Kill Your Baby*, 2–3; and Golden, *Social History of Wet Nursing*, 37. For a broader perspective on these changes, see, for example, Paul Starr, *The Social Transformation of American Medicine: The Rise of a Sovereign Profession and the Making of a Vast Industry* (New York: Basic Books, 1982), part 1; Ruth Schwartz Cowan, *More Work for Mother: The Ironies of Household Technology from the Open Hearth to the Microwave* (New York: Basic Books, 1983), 69–101; Alice Kessler-Harris, *In Pursuit of Equity: Women, Men and the Quest for Economic Citizenship* (New York: Oxford University Press, 2001), 3–18; Kathy Peiss, *Cheap Amusements: Working Women and Leisure in Turn-of-the-Century New York* (Philadelphia: Temple University Press, 1986), 34–55; Regina G. Kunzel, *Fallen Women, Problem Girls: Unmarried Mothers and the Professionalization of Social Work, 1890–1945* (New Haven, CT: Yale University Press, 1993), 1–8; and Amy Bentley, *Eating for Victory: Food Rationing and the Politics of Domesticity* (Urbana: University of Illinois Press, 1998), 114–41.

13. Apple, *Mothers and Medicine*, 166; Wolf, *Don't Kill Your Baby*, 3.

14. Pam Carter, *Feminism, Breasts and Breast Feeding* (New York: St. Martin's Press, 1995); Naomi Baumslag and Dia L. Michels, *Milk, Money, and Madness: The Culture and Politics of Breastfeeding* (Westport, CT: Bergin and Garvey, 1995); Barbara L. Behrmann, *The Breastfeeding Café: Mothers Share the Joys, Challenges, and Secrets of Nursing* (Ann Arbor: University of Michigan Press, 2005); Valerie Fildes, *Breasts, Bottles and Babies: A History of Infant Feeding* (Edinburgh: Edinburgh University Press, 1986); Bernice Hausman, *Mother's Milk: Breastfeeding Controversies in American Culture* (New York: Routledge, 2003).

15. Apple, *Mothers and Medicine*, 165; Jacqueline H. Wolf, "What Feminists Can Do for Breastfeeding and What Breastfeeding Can Do for Feminists," *Signs* 31, no. 2 (2006): 397–424, quotation on 399. Wolf states that initiation rates are "largely meaningless," because they capture any infant fed even once at the breast. I argue that these initiation rates can and do in fact tell us about women's continued interest in breastfeeding.

16. Apple, *Mothers and Medicine*, 19, 157.

17. Ibid., 97.

18. Ibid., 165.

19. Golden, *Social History of Wet Nursing*, 37; Wolf, *Don't Kill Your Baby*, 2–3.

20. Apple, *Mothers and Medicine*, 175. Unless otherwise noted, I use "bottle" and "formula" interchangeably, to incorporate all forms of non-breast-milk mixtures that mothers have fed their infants with medical advice throughout the majority of the twentieth century.

21. Golden, *Social History of Wet Nursing*, 156–58; Wolf, *Don't Kill Your Baby*, 86–87.

22. Wolf, *Don't Kill Your Baby*, 3, 10.

23. Golden, *Social History of Wet Nursing*, 198–200.

24. Swanson, "Human Milk as Technology," 21–37.

25. Apple, *Mothers and Medicine*, 177–78; Wolf, "What Feminists Can Do," 407–9.

26. Christina Bobel, "Bounded Liberation: A Focused Study of La Leche League International," *Gender and Society* 15, no. 1 (February 2001): 130–51; Linda Blum, *At the Breast: Ideologies of Breastfeeding and Motherhood in the Contemporary United States* (Boston: Beacon Press, 1999), 63–107; Lynn Y. Weiner, "Reconstructing Motherhood: The La Leche League in Postwar America," *Journal of American History* 80, no. 4 (March 1994): 1357–81; Jule DeJager Ward, *La Leche League: At the Crossroads of Medicine, Feminism, and Religion* (Chapel Hill: University of North Carolina Press, 2000), 165–68.

27. *The Womanly Art of Breastfeeding* (Franklin Park, IL: La Leche League, 1963 [orig. 1958]).

28. Ward, *La Leche League*, 3–5.

29. Weiner, "Reconstructing Motherhood," 1358.

30. League leaders and members participated in an active and growing network of women's health organizations.

31. Rima D. Apple, *Perfect Motherhood: Science and Childrearing in America* (New Brunswick, NJ: Rutgers Press, 2006), 136; Tompson to Montagu, 14 December 1960, Marian Tompson Papers, LLLIR; Mary Ann Cahill, ed., *Seven Voices, One Dream* (Schaumburg, IL: La Leche League International, 2001), 5–11. Letters and the oral histories of several of the original founders reveal that four of the seven original LLL founders experienced natural and home births in the 1940s and 1950s. Dr. Gregory White, husband of an LLL founder, was instrumental in offering natural and home births as well as breastfeeding support to his community.

32. "Women in Health Care and Pediatrics," *Pediatrics* 71, no. 4 (April 1983): 681–87.

33. Katherine Bain, "The Incidence of Breast Feeding in Hospitals in the United States," *Pediatrics* 2, no. 3 (1948): 313–20, 315.

34. O. William Anderson, "Infant Feeding around the World: Breast Feeding in the Culture of the United States," *Quarterly Review of Pediatrics* 13, no. 4 (1958): 203–11, 204.

35. Apple, *Mothers and Medicine*, 16–19; Julia Grant, *Raising Baby by the Book* (New Haven, CT: Yale University Press, 1998), 1–10; Ann Hulbert, *Raising America: Experts, Parents, and a Century of Advice about Children* (New York: Alfred A. Knopf, 2003), 7–10.

36. See, for example, Meghan Crnic and Cynthia Connolly, "'They Can't

Help Getting Well Here': Seaside Hospitals for Children in the United States, 1872–1917," *Journal of the History of Childhood and Youth* 2 (Spring 2009): 220–33; Julia Guarneri, "Changing Strategies for Child Welfare, Enduring Beliefs about Childhood: The Fresh Air Fund, 1877–1926," *Journal of the Gilded Age and Progressive Era* 11, no. 1 (January 2012): 27–70; and Daniel Freund, *American Sunshine: Diseases of Darkness and the Quest for Natural Light* (Chicago: University of Chicago Press, 2012).

37. Kathleen Jones, *Taming the Troublesome Child: American Families, Child Guidance, and the Limits of Psychiatric Authority* (Cambridge, MA: Harvard University Press, 1999), 4; Alice Boardman Smuts, *Science in the Service of Children, 1893–1935* (New Haven, CT: Yale University Press, 2008), 103.

38. Apple, *Mothers and Medicine*, 180; Blum, *At the Breast*, 161–72; Hausman, *Mother's Milk*, 26, 219–25. A handful of studies indicate that between 1930 and 1955, there was a noticeable reversal in the relationship between breastfeeding and socioeconomic status. In 1930, a study found that lower-class women were more likely to breastfeed. By the late 1940s, rates appeared even across class, and by 1955, middle-class women married to college-educated men were most likely to breastfeed.

39. Smuts, *Science in the Service*, 53.

40. L. Emmett Holt, *The Care and Feeding of Children*, 5th ed. (New York: D. Appleton, 1909 [orig. 1894]), 165, 168.

41. John B. Watson, *Psychological Care of Infant and Child* (New York: W. W. Norton, 1928), 69, 85, 87.

42. Nikolas Rose, *Inventing Our Selves: Psychology, Power, and Personhood* (New York: Cambridge University Press, 1998), 1–3.

43. Marga Vicedo, "The Evolution of Harry Harlow," *History of Psychiatry* 21, no. 1 (2010): 1–16; Vicedo, "The Social Nature of the Mother's Tie to Her Child," *British Society for the History of Science* 44, no. 3 (2011): 401–26. Vicedo's work has done much to illuminate this history of the science of instinct. See also Marga Vicedo, *The Nature and Nurture of Love: From Imprinting to Attachment in Cold War America* (Chicago: University of Chicago Press, 2013).

44. Media coverage of these scientific findings reflects Americans' willingness to accept implications of scientific studies on their own lives. See, for example, "Warping of Child Traced to Family," *New York Times* (12 April 1952), 16.

45. Weiner, "Reconstructing Motherhood," 1358; Grant, *Raising Baby*, 209. Weiner observes that the League "emphasized naturalism" over scientific motherhood and Grant also notes a more general turn toward "naturalism" in this era.

46. Joan J. Mathews and Kathleen Zadak, "The Alternative Birth Movement in the United States: History and Current Status," *Women and Health* 17, no. 1 (1991): 39–56; Weiner, "Reconstructing Motherhood," 1366.

47. Margarete Sandelowski, *Pain, Pleasure, and American Childbirth: From*

the Twilight Sleep to the Read Method, 1914–1960 (Westport, CT: Greenwood Press, 1984), xi–xii.

48. Judith W. Leavitt, *Brought to Bed: Childbearing in America, 1750–1950* (New York: Oxford University Press, 1988), 171; Edith B. Jackson, "Should Mother and Baby Room Together?" *American Journal of Nursing* 46, no. 1 (January 1946): 17–19; Sandelowski, *Pain, Pleasure, and American Childbirth*, 85.

49. Jacqueline H. Wolf, *Deliver Me from Pain: Anesthesia and Birth in America* (Baltimore: Johns Hopkins University Press, 2009), 105–6.

50. Hazel Corbin to Mrs. Kelliher, 12 January 1955, Folder 8, Box 15, Series V, Boston Association for Childbirth Education Records (hereafter BACE), Schlesinger Library Special Collections, Radcliffe Institute for Advanced Study (hereafter SLSC). See also Sonya Michel, "American Women and the Discourse of the Democratic Family in WWII," in *Behind the Lines: Gender and the Two World Wars* (New Haven, CT: Yale University Press, 1987), 167.

51. Rev. Henry V. Sattler to Kelliher, 5 March 1958, Folder 8, Box 15, Series V, BACE Records, SLSC. Catholic physicians and laity played a central, and largely underexplored, role in the movements of natural childbirth and breastfeeding during this era.

52. "ICEA Instructor's Kit," 1962, Folder 17, Box 15, Series V, BACE Records, SLSC.

53. Wolf, *Deliver Me from Pain*, 140. This is not to imply that Read's method "worked" for all women, all of the time. Wendy Kline has documented the experiences of "failed" natural childbirth for women who were seeking empowering births. Wendy Kline, *Bodies of Knowledge: Sexuality, Reproduction and Women's Health in the Second Wave* (Chicago: University of Chicago Press, 127–30).

54. "About Membership in the International Childbirth Education Association," Folder 12, Box 15, Series V, BACE Records, SLSC.

55. Ibid.

56. Wolf, *Deliver Me from Pain*, 139.

57. Weiner, "Reconstructing Motherhood," 1359.

58. For a nuanced discussion of how medical knowledge can both constrain the parameters of the patient experience and also serve as a tool of patient empowerment, see Charles Rosenberg, "Explaining Epidemics and Other Studies in the History of Medicine," in *Explaining Epidemics and Other Studies in the History of Medicine* (New York: Cambridge University Press, 1992), 32–56.

59. I have adopted Rebecca Jo Plant's terminology to describe these two constructs. See Rebecca Jo Plant, *Mom: The Transformation of Motherhood in Modern America* (Chicago: University of Chicago Press, 2010), 2, 6–7.

60. Important works on Victorian motherhood include Ruth H. Bloch, "American Feminine Ideals in Transition: The Rise of the Moral Mother, 1790–1815," *Feminist Studies* 4 (June 1978): 100–126; Nancy F. Cott, *The Bonds of Womanhood: "Woman's Sphere" in New England, 1780–1835* (New Haven,

CT: Yale University Press, 1977); Ann Douglas, *The Feminization of American Culture* (New York: Alfred A. Knopf, 1979); Sylvia D. Hoffert, *Private Matters: American Attitudes toward Childbearing and Infant Nurture in the Urban North, 1800–1860* (Urbana: University of Illinois Press, 1989); Amy Kaplan, "Manifest Domesticity," *American Literature* 70, no. 3 (September 1998): 581–606; Jan Lewis, "Mother's Love: The Construction of Emotion in Nineteenth-Century America," in *Social History and Issues in Human Consciousness: Some Interdisciplinary Connections*, ed. Andrew E. Barnes and Peter N. Stearns (New York: New York University Press, 1989), 209–29; Mary P. Ryan, *Cradle of the Middle Class: The Family in Oneida County, New York, 1790–1865* (Cambridge: Cambridge University Press, 1981); and Ryan, *The Empire of the Mother: American Writing about Domesticity* (New York: Harrington Park Press, 1982).

61. See, for example, Robyn Muncy, *Creating a Female Dominion in American Reform, 1890–1935* (New York: Oxford University Press, 1991).

62. See, for example, Grant, *Raising Baby*, 5; Leavitt, *Brought to Bed*, 4–5; Morantz-Sanchez, *Sympathy and Science*, 196–97.

63. Plant, *Mom*, 6.

64. Weiner, "Reconstructing Motherhood," 1381; Wolf, "What Feminists Can Do," 409.

65. Plant, *Mom*, 15–16.

66. Ibid., 3.

67. Vicedo, *Nature and Nurture of Love*, 16.

68. Linda Nash, *Inescapable Ecologies: A History of Environment, Disease, and Knowledge* (Berkeley: University of California, 2006), 5. See also Nancy Langston, *Toxic Bodies: Hormone Disruptors and the Legacy of DES* (New Haven, CT: Yale University Press, 2010); Gregg Mitman, *Breathing Space: How Allergies Shape Our Lives and Landscapes* (New Haven, CT: Yale University Press, 2007); and Christopher Sellers, *Crabgrass Crucible: Suburban Nature and the Rise of Environmentalism in Twentieth-Century America* (Chapel Hill: University of North Carolina Press, 2012).

69. Edwin P. Laug, Frieda M. Kunze, and C. S. Prickett, "Occurrence of DDT in Human Fat and Milk," *Archives of Industrial Hygiene* 3 (March 1951): 245–46.

70. Londa Schiebinger, "Why Mammals Are Called Mammals: Gender Politics in Eighteenth-Century Natural History," *American Historical Review* 98, no. 2 (April 1993): 382–411.

71. Rachel Carson, *Silent Spring* (Boston: Houghton Mifflin Press, 1962); Sandra Steingraber, *Living Downstream: An Ecologist's Investigation of Cancer and the Environment* (Cambridge, MA: Da Capo Press, 2010); Terry Tempest Williams, *Refuge: An Unnatural History of Family and Place* (New York: Pantheon Books, 1991).

72. Sellers, *Crabgrass Crucible*, 294–96.

73. Linda Blum, *At the Breast: Ideologies of Breastfeeding and Motherhood*

in the Contemporary United States (Boston: Beacon Press, 1999), 171; Golden, *Social History of Wet Nursing*, 73; Stephanie Jones-Rogers, "Black Milk: Maternal Bodies, Wet Nursing and the Value of Black Women's Invisible Labor in the Antebellum Slave Market" (paper presented at the Berkshire Conference on the History of Women, Amherst, Massachusetts, 9–12 June 2011).

74. "More Slavery at the South," *Independent* (25 January 1912), 196–200, republished online at http://historymatters.gmu.edu/d/80 (accessed 7 August 2014).

75. Gail Bederman, *Manliness and Civilization: A Cultural History of Gender and Race in the United States, 1880–1917* (Chicago: University of Chicago Press, 1995), 95–101; Stephen Jay Gould, *The Mismeasure of Man* (New York: W. W. Norton, 1996), 142–51; Marie Jenkins Schwartz, *Birthing a Slave: Motherhood and Medicine in the Antebellum South* (Cambridge, MA: Harvard University Press, 2006), 252; Leila Zenderland, *Measuring Minds: Henry Herbert Goddard and the Origins of American Intelligence Testing* (Cambridge: Cambridge University Press, 2001), 315–18; Keith Wailoo, *Dying in the City of the Blues: Sickle Cell Anemia and the Politics of Race and Health* (Chapel Hill: University of North Carolina Press, 2001).

76. Bederman, *Manliness and Civilization*, 1–5, 29, 36; Donna Harraway, "Teddy Bear Patriarchy: Taxidermy in the Garden of Eden, New York City, 1908–36," *Social Text* 11 (Winter 1984–85): 20–64. The participation by white women in maintaining and exploiting these racial hierarchies of motherhood has a long history, some of the best analyses of these dynamics are in Laura Wexler, *Tender Violence: Domestic Visions in an Age of U.S. Imperialism* (Chapel Hill: University of North Carolina Press, 2000),11–13, 52–93, 178–80.

77. Blum, *At the Breast*, 47, 169.

78. For more on recent scholarship that addresses the issues of race, structural constraints and the "choice" to breastfeed, see Paige Hall Smith, Bernice Hausman, and Miriam Labbock, eds., *Beyond Health, Beyond Choice: Breastfeeding Constraints and Realities* (New Brunswick, NJ: Rutgers University Press, 2012).

Chapter 1

1. Watson, *Psychological Care of Infant and Child*, front matter.

2. Read first published his work on natural childbirth in 1933 in a book titled *Natural Childbirth*. In 1942, he published another book under the title *Revelation of Childbirth*, and in 1944, he republished it in the United States under the title *Childbirth without Fear*, which is the book he is perhaps best known for authoring. Grantly Dick-Read, *Childbirth without Fear* (New York: Harper and Brothers, 1953 [orig. 1944]); Mary Thomas, ed., *Post-War Mothers: Childbirth Letters to Grantly Dick-Read, 1946–1956* (Rochester, NY: Univer-

sity of Rochester Press, 1997), ix; Sandelowski, *Pain, Pleasure and American Childbirth*, 55.

3. On "scientific motherhood," see Apple, *Mothers and Medicine*; Grant, *Raising Baby by the Book*; Molly Ladd-Taylor, *Mother-Work: Women, Child Welfare, and the State, 1890–1930* (Urbana: University of Illinois Press, 1994); Ladd-Taylor, *Raising a Baby the Government Way: Mothers' Letters to the Children's Bureau, 1915–1932* (New Brunswick, NJ: Rutgers University Press, 1986), 1–46; and Barbara Ehrenreich and Deirdre English, *For Her Own Good: 150 Years of Experts' Advice to Women* (New York: Anchor Books, 2005 [orig. 1978]), 201–30. On recent arguments about the importance of emotion in the social sciences during this period, see Vicedo, *Nature and Nurture of Love*, 17–21.

4. Rose, *Inventing Our Selves*, 11. Rose suggests the terminology of the "psy" disciplines to refer to the transportability of psychology, not as "a body of abstracted theories and explanations," but as an "intellectual technology," a "way of making visible and intelligible certain features of persons, their conducts, and their relations with one another." I utilize this terminology in my work to refer to the interdisciplinary movement of ideas and tools that mid-twentieth-century "psy"-entists utilized to make the nature of motherhood, and breastfeeding, "intelligible."

5. Watson, *Psychological Care*, 7.

6. Apple, *Mothers and Medicine*, 97.

7. Watson, *Psychological Care*, 11.

8. Ladd-Taylor and Umansky, *Bad Mothers*, 1–11.

9. C. Anderson Aldrich, *Babies Are Human Beings: An Interpretation of Growth* (New York: Macmillan, 1938), 70–71.

10. "Margaret Ribble, Psychoanalyst, 80," *New York Times* (21 July 1971), 38; Margarethe A. Ribble [*sic*], "The Significance of Infantile Sucking for the Psychic Development of the Individual," *Journal of Nervous and Mental Disease* 90, no. 4 (October 1939): 455–63.

11. Margaret Ribble, *The Rights of Infants: Early Psychological Needs and Their Satisfaction* (New York: Columbia University Press, 1944 [orig. 1943]), 3.

12. Ibid.

13. Ibid., 8.

14. Ibid., 33.

15. Ibid., 33–34.

16. M. C. Woodard, "Breast or Bottle for a Baby?" *Parents' Magazine* 20 (August 1945), 20–21, 78, 80, 82, 84.

17. Michael E. Lamb, Ross A. Thompson, William Gardner, and Eric L. Charnov, *Infant-Mother Attachment: The Origins and Developmental Significance of Individual Differences in Strange Situation Behavior* (Hillsdale, NJ: Lawrence Erlbaum Associates, 1985), 8. See also Jean-Jacques Rousseau, *Emile, or, On Education*, trans. Allan Bloom (New York: Basic Books, 1979).

18. David Levy, "Fingersucking and Accessory Movements in Early Infancy: An Etiologic Study," *American Journal of Psychiatry* 84, no. 6 (May 1928): 881–918, quotation on 881.

19. Ibid., 883.

20. Raymond Dyer, *Her Father's Daughter: The Work of Anna Freud* (New York: Jason Aronson, 1983), 239.

21. Ibid., 240.

22. Melanie Klein, "The Psychoanalytic Play Technique: Its History and Significance," in *The Selected Melanie Klein*, ed. Juliet Mitchell (London: Hogarth Press, 1986), 52. Originally printed in *American Journal of Orthopsychiatry* 25, no. 2 (April 1955): 223–37.

23. Meira Likierman, *Melanie Klein: Her Work in Context* (New York: Continuum Press, 2005 [orig. 2001]), 55.

24. Ibid., 81.

25. See, for example, Melanie Klein, "The Study of Gratitude," in *The Selected Melanie Klein*, ed. Juliet Mitchell (London: Hogarth Press, 1986 [orig. 1956]), 211.

26. Levy, "Fingersucking and Accessory Movements," 917.

27. Ribble, "Significance of Infantile Sucking," 455.

28. Ibid., 457.

29. See Margarethe A. Ribble [*sic*], "Clinical Studies of Instinctive Reactions in New Born Babies," *American Journal of Psychiatry* 95, no. 1 (1 July 1938): 149–60, see especially "discussion" on 158–60; Samuel R. Pinneau, "A Critique on the Articles by Margaret Ribble," *Child Development* 2, no. 4 (December 1950): 203–28; and René Spitz, "Reply to Dr. Pinneau," *Psychological Bulletin* 52, no. 5 (September 1955): 453–59. Marga Vicedo's work has done much to integrate Ribble into the history of motherhood studies. See Vicedo, *Nature and Nurture of Love*, 82–83.

30. "Margaret Ribble, Psychoanalyst, 80," *New York Times* (21 July 1971), 38.

31. "Margaret Ribble, Psychiatrist, Dies in Warrenton," *Free Lance-Star* (19 July 1971), 7.

32. Margaret Ribble to Anna Freud, 20 December 1951, Folder 7, Box 86, Margaret A. Ribble, 1950–55, Anna Freud Papers, Library of Congress, MSS-49700, 86/7.

33. Margaret Ribble to Anna Freud, 23 July 1951, Folder 7, Box 86, Margaret A. Ribble, 1950–1955, Anna Freud Papers, Library of Congress, MSS-49700, 86/7.

34. Marga Vicedo, "The Father of Ethology and the Foster Mother of Ducks: Konrad Lorenz as Expert on Motherhood," *Isis* 100 (June 2009): 273.

35. René A. Spitz, *Psychogenic Disease in Infancy* [film] (1952), Prelinger Archive, available online at https://archive.org/details/PsychogenicD (accessed 17 August 2014).

36. René A. Spitz, *The First Year of Life: A Psychoanalytic Study of Normal and Deviant Development of Object Relations* (New York: International Universities Press, 1965).

37. Ivan C. Berlien, "Growth as Related to Mental Health," *American Journal of Nursing* 56, no. 9 (September 1956): 1142–45, quotation on 1143.

38. Ibid.

39. John Holmes, *John Bowlby and Attachment Theory* (New York: Routledge, 1993), 62. This version of events, however, has recently been challenged by work done by historian of science Marga Vicedo. Through detailed analysis of Lorenz and his work, Vicedo has offered a persuasive account of Lorenz as a eugenicist who, from the beginning of his career, drew untested parallels between his work on birds and the human experience. See Vicedo, "Father of Ethology and the Foster Mother of Ducks," 264.

40. Holmes, *John Bowlby*, 64.

41. Mary Ainsworth and John Bowlby, "An Ethological Approach to Personality Development," *American Psychologist* 46, no. 4 (1991): 331–41, 335. Ainsworth and Bowlby coauthored this paper and delivered it as the Distinguished Scientific Contributions Award Address at the Ninety-Eighth Annual Convention of the American Psychological Association in August 1990. Less than one month later, John Bowlby passed away. Ainsworth finished preparing the article and had it published, as Bowlby had intended to do himself.

42. Ibid., 336.

43. John Bowlby, *Attachment and Loss: Attachment*, vol. 1 (New York: Basic Books, 1969), 183.

44. John Bowlby, *Maternal Care and Mental Health* (Geneva: World Health Organization, 1951), 7. Originally published in the *Bulletin of the World Health Organization*, no. 3 (1951): 355–534.

45. Bowlby, *Maternal Care*, 11.

46. Ibid.

47. Ibid.

48. Ibid.

49. Ibid., 15.

50. Holmes, *John Bowlby*, 27.

51. Frances C. Bauer, "Mother Shouldn't Smother," *New York Times* (28 July 1963), 184. The *New York Times* tracked Bowlby's work, beginning with a 1949 announcement of his study of infants at the Tavistock Clinic in London. See "Child Separation Study: British Psychiatrists Observe Effects of Missing Mothers," *New York Times* (10 April 1949), 48; and "Warping of Child Traced to Family: World Health Study Suggests Social Aids to Correct Lack of Maternal Care," *New York Times* (12 April 1952), 16. The World Health Organization sponsored a follow-up study to Bowlby's report in 1962. See Mary D. Ainsworth, R. G. Andry, Robert G. Harlow, S. Lebovici, Margaret Mead, Dane G. Prugh,

and Barbara Wootton, *Deprivation of Maternal Care: A Reassessment of Its Effects*, Public Health Papers No. 14 (Geneva: World Health Organization, 1962).

52. Bauer, "Mother Shouldn't Smother," 184.

53. John Bowlby, "The Nature of the Child's Tie to His Mother," *International Journal of Psycho-Analysis* 39 (1958): 350–73, quotation on 62.

54. Bowlby, *Maternal Care*, 20.

55. Ibid., 15.

56. Ibid.

57. Ibid.

58. John Bowlby to Harry Harlow, 8 August 1957, from the personal collection of F. C. P. van der Horst, Faculty of Social and Behavioural Sciences at the Centre for Child and Family Studies, Leiden University, the Netherlands (hereafter FCPH).

59. Harry F. Harlow, "The Nature of Love," *American Psychologist* 13, no. 12 (1958): 673–85. Originally delivered as the Presidential Address at the 66th Annual Convention of the American Psychological Association, Washington, DC, 31 August 1958.

60. Ibid., 677.

61. Ann Roe's interview with Harry Harlow, Harry F. Harlow, 1905–1982, Ann Roe Papers, 1904–1991, American Philosophical Society, Philadelphia (hereafter APS).

62. Harry Harlow to John Bowlby, 3 October 1957, FCPH.

63. On the shift toward an embodied concept of identity in American womanhood, see Joan Jacobs Brumberg, *The Body Project: An Intimate History of American Girls* (New York: Vintage Books, 1998 [orig. 1997]); and R. Marie Griffith, *Flesh and Spirit in American Christianity* (Berkeley: University of California Press, 2004).

64. See, for example, Lee D. Baker, "Franz Boas out of the Ivory Tower," *Anthropological Theory* 4, no. 1 (2004): 29–51; Regna Darnell, "Re-envisioning Boas and Boasian Anthropology," *American Anthropologist* 102, no. 4 (2000): 896–900; and Douglas Cole, *Franz Boas: The Early Years, 1858–1906* (Seattle: University of Washington Press, 1999).

65. Bruce Dakowski, *Margaret Mead: Coming of Age* [film], produced and directed by André Singer (Central Independent Television, 1990). For more information on Mead's *Redbook* columns, see Margaret Mead, *Some Personal Views*, ed. Rhoda Metraux (New York: Walker and Company, 1979), a collection of Mead's popular columns. Copies of Mead's films are available at the Museum Library located at the University Museum of Anthropology and Archaeology at the University of Pennsylvania, Philadelphia.

66. Lois W. Banner, *Intertwined Lives: Margaret Mead, Ruth Benedict, and Their Circle* (New York: Vintage, 2004), 364.

67. Jane Howard, *Margaret Mead: A Life* (New York: Simon and Schuster, 1984), 243.

68. Banner, *Intertwined Lives*, 266; Margaret W. Rossiter, *Women Scientists in America: Before Affirmative Action, 1940–1972*, vol. 2 (Baltimore: Johns Hopkins University Press, 1995), 244, 313–14.

69. Grant, *Raising Baby*, 171, 184; May, *Homeward Bound*, 10–11, 116–17.

70. Margaret Mead and Gregory Bateson, *Childhood Rivalry in Bali and New Guinea* [film], narrated by M. Mead, from the Character Formation in Different Cultures Series (1952); Margaret Mead and Gregory Bateson, *First Days in the Life of a New Guinea Baby* [film], narrated by M. Mead, from the Character Formation in Different Cultures Series (1952).

71. Margaret Mead, *Male and Female: A Study of the Sexes in a Changing World* (London: Gollancz, 1950), 268–70.

72. Jeannie J. Sakol, "Remarkable Woman: Margaret Mead," *McCall's* 97, no. 6 (June 1970), 80–81, 81.

73. "Interview with Margaret Mead," *Family Health* 10 (October 1978), 48–53, 50.

74. Margaret Mead, "Margaret Mead Answers," *Redbook* 132 (January 1969), 33–35, 33.

75. Rossiter, *Women Scientists in America*, vol. 2, 33. Rossiter refers to the 1950s and 1960s as an era characterized by the "remasculinization of science." This was no less true of studies of motherhood, childbirth, and child rearing at the time, even though these areas of scientific inquiry were relatively more likely to have female PhDs than other fields.

76. Philip Morehouse McGarr and Fanny Scott Rumely, "The Autobiography of Dr. Edward A. Rumely: The Formative Years, 1882–1900," *Indiana Magazine of History* 66, no. 1 (March 1970), 1–39. Note 1 contains information about Mrs. Rumely and the historical collections of the Scott and Rumely ancestors.

77. Rumely Family Christmas Letter, December 1948, Folder 7, Box 3, Series VI, Edward A. Rumely Personal Papers, Special Collections and University Archives, University of Oregon, Eugene, Oregon (hereafter EARPP).

78. Rumely Family Christmas Letter, December 1947, Folder 7, Box 3, Series VI, EARPP.

79. "Mind and Milk," *Time* 55, no. 25 (19 June 1950), 83.

80. Michael M. Newton and Niles R. Newton, "The Let-Down Reflex in Human Lactation," *Journal of Pediatrics* 33, no. 6 (1948): 698–704. The article itself gives no hint as to the identity of the research subject, but in his write-up on the Newtons' work, science journalist Edward Brecher identified the mother in the experiments as Niles Newton herself. See Edward M. Brecher, *The Sex Researchers* (New York: Little, Brown, 1969), 174.

81. As Michael's own career trajectory suggests, however, his primary inter-

ests were in the study and treatment of gynecological cancers. It seems it was Niles who drove their coauthored work on breastfeeding.

82. Newton, *Maternal Emotions*.

83. Margaret Mead and Niles R. Newton, "Conception, Pregnancy, Labor and the Puerperium in Cultural Perspective," in *Proceedings of the First International Congress of Psychosomatic Medicine and Childbirth* (Paris: Gauthier-Villars, 1965), 54.

84. Niles R. Newton, "Report from Paris: The International Congress on Psychosomatic Obstetrics," *Child and Family Digest* (October 1962), Correspondence Files—Niles Newton, LLLIR. Newton also exchanged numerous letters with Mead about this conference, including a letter dated July 1962 in which she gave Mead a detailed rundown of the Congress's events.

85. Margaret Mead and Niles R. Newton, "Cultural Patterning of Perinatal Behavior," in *Childbearing: Its Social and Psychological Aspects*, ed. S. A. Richardson and A. F. Guttmacher (Baltimore: Williams and Wilkins, 1967), 181.

86. Ibid.

87. Ibid., 186.

88. Ibid., 182.

89. *Womanly Art* (1963), 141–46.

90. Niles Newton, "Breast Feeding Today," presented at the National Convention for Childbirth Education at Milwaukee, Wisconsin, 21 May 1960, Correspondence Files—Niles Newton, LLLIR.

91. Froehlich to Newton, 19 April 1967, Correspondence Files—Niles Newton, LLLIR.

92. Harlow, "Nature of Love," 685.

93. Newton, *Maternal Emotions*, 50.

94. Plant, *Mom*, 13.

95. Robert R. Sears, "Personality Development in Contemporary Culture," *Proceedings of the American Philosophical Society* 92, no. 5 (1948), 363–70.

Chapter 2

1. Interview with Barbara (1 October 2008), author's files.

2. Ibid.

3. Apple, *Mothers and Medicine*, 36, 165–66; Wolf, *Don't Kill Your* Baby, 2–7.

4. This phenomenon has also been explored at length in Richard A. Meckel, *Save the Babies: American Public Health Reform and the Prevention of Infant Mortality, 1850–1929* (Ann Arbor: University of Michigan Press, 1998 [orig. Johns Hopkins University Press, 1990]), 52. In the epilogue of her 2001 monograph, *Don't Kill Your Baby*, Wolf explains that by the 1930s a new generation of physicians came into practice, most having never seen a baby die from

contaminated cow's milk and having never seen a mother breastfeeding. Wolf suggests that physicians' ignorance about lactation in this period was partly to blame for breastfeeding's continued decline, along with a continuation of practices of "mixed feeding" (relying on a combination of breastfeeding and artificial feeding), early weaning, the premature introduction of solid foods, and rigid and dehumanizing hospital maternity and nursery routines. I disagree with the argument that by the 1920s and 1930s *most* women were choosing not to attempt breastfeeding. I also do not believe that this signaled the beginning of a long period in the mid-twentieth century when mothers did not *want* to breastfeed. The studies and medical reports that I have looked at in the 1930s, 1940s, 1950s, and 1960s do not support this claim outright. Rather than discount breastfeeding whenever it accompanied additional feeding practices such as table scraps and cow's milk formulas or when it lasted for only a brief time, I have chosen to follow the persistence of women's choices to breastfeed under the hypothesis that women who attempted breastfeeding, even for just a few days or weeks, did so because they believed it had value, and often they did so despite the fact that most hospitals implemented policies by midcentury that actively discouraged breastfeeding. See Wolf, *Don't Kill Your Baby*, 189–96.

5. Apple, *Mothers and Medicine*, 131.

6. Nutrition Committee of the Canadian Paediatric Society and the Committee on Nutrition of the American Academy of Pediatrics, "Breast Feeding: A Commentary in Celebration of the International Year of the Child, 1979," *Pediatrics* 62, no. 4 (October 1978): 591–601, quotation on 597.

7. Clement A. Smith, C. Heaton, J. Y. Harshberger, Benjamin Spock, I. T. Nathanson, "Present Day Attitudes towards Breast Feeding," *Pediatrics* 6, no. 4 (October 1950): 656–59.

8. Bain, "Incidence of Breast Feeding," 313–20.

9. Smith et al., "Present Day Attitudes," 657.

10. Ibid., 656.

11. Clifford G. Grulee, Heyworth N. Sanford, and Paul H. Herron, "Breast and Artificial Feeding Influence on Morbidity and Mortality of 20,000 infants," *Journal of the American Medical Association* 103, no. 10 (8 September 1934): 735–39, quotation on 738.

12. Clifford G. Grulee, Heyworth N. Sanford, and Paul H. Herron, "Influence of Breast and Artificial Feeding on Morbidity and Mortality on 20,000 Infants," *Journal of Pediatrics* 9 (August 1936): 223–25.

13. Ibid.

14. Howard J. Morrison, "Breast Feeding," *Pediatrics* 1, no. 12 (December 1950): 1473–82.

15. S. J. Cowell, "The Feeding of Normal and Premature Infants," *Proceedings of the Nutrition Society*, Seventieth Scientific Meeting, London School of Hygiene and Tropical Medicine (6 October 1951), 207–29; Ernest H. Watson,

"Breast Feeding of Normal Infants," *GP* 4, no. 5 (November 1951): 53–58; Manuel M. Glazier, "Comparing the Breast and Bottle Fed Infants," *New England Journal of Medicine* 203, no. 13 (25 September 1930): 626–31.

16. Harold Waller, "The Early Failure of Breast Feeding: A Clinical Study of Its Causes and Their Prevention," *Archives of Disease in Childhood* 21 (March 1946): 1–12, quotation on 1.

17. Frank Howard Richardson, "Breast Feeding Comes of Age," *Journal of the American Medical Association* 142, no. 12 (25 March 1950): 863–67, quotation on 863.

18. Ibid.

19. Julius Parker Sedgwick, "A Preliminary Report of the Study of Breast Feeding in Minneapolis," *American Journal of Diseases of Children* 21 (May 1921): 455–64, see 459. Sedgwick included infants who received mixed feedings of both breast milk and formula in his numbers. When broken down into infants fed only breast milk, his numbers are slightly less impressive—though still higher than most physicians could report by the 1940s.

20. Richardson, "Breast Feeding," 863.

21. Ibid.

22. M. L. Turner, "Some Observations from Nature," *Transactions from the Section on Diseases of Children at the American Medical Association* (1923), 63, quoted in Richardson, "Breast Feeding," 863.

23. Stuart Shelton Stevenson, "The Adequacy of Artificial Feeding in Infancy," *Journal of Pediatrics* 31, no. 6 (December 1947): 616–30, see 621, 623.

24. Niles R. Newton and Michael Newton, "Recent Trends in Breast Feeding: A Review," *American Journal of the Medical Sciences* 221, no. 6 (June 1951): 691–99, see 692.

25. Ibid.

26. Paul Gyorgy, "A Hitherto Unrecognized Biochemical Difference between Human Milk and Cow's Milk," *Pediatrics* 11, no. 2 (February 1953): 98–108, see 105.

27. Andrew C. McCandlish, "The Possibility of Increasing Milk and Butterfat Production by the Administration of Drugs," *Journal of Dairy Science* 1, no. 6 (1918): 475–86; O. Zietzschmann, "Étude sur les vaches qui 'retiennent' leur lait," *Le Lait* 2, no. 4 (1922): 229–37.

28. W. L. Gaines, "A Contribution to the Physiology of Lactation," *American Journal of Physiology* 38, no. 2 (1915): 285–312.

29. Walter B. Cannon and E. M. Bright, "A Belated Effect of Sympathectomy on Lactation," *American Journal of Physiology* 97, no. 2 (1931): 319–21.

30. K. McKenzie, "An Experimental Investigation of the Mechanism of Milk Secretion," *Journal of Experimental Physiology* 4 (January 1911): 305–30.

31. P. Ingelbrecht, "L'influence du système nerveux central sur la mamelle lactante chez le rat blanc," *Comptes Rendus Hebdomadaires des Séances et Mé-*

moires de la Société de Biologie et des Ses Filiales et Associées 120 (1935): 1369–71. Cited in Fordyce Ely and W. E. Petersen, "Factors Involved in the Ejection of Milk," *Journal of Dairy Science* 24, no. 3 (March 1941): 211–23.

32. E. A. Schafer, "On the Effect of Pituitary and Corpus Luteum Extracts on the Mammary Gland in the Human Subject," *Quarterly Journal of Experimental Physiology* 6, no. 1 (1913): 17–19.

33. Ely and Petersen, "Factors Involved," 211.

34. Diane Wiessinger, Diana West, and Teresa Pitman, *The Womanly Art of Breastfeeding*, La Leche League International, 8th ed. (New York: Ballantine Books, 2010), 23–24.

35. On the study of physiology and hormone research, see Sheila M. Rothman and David J. Rothman, *The Pursuit of Perfection: The Promise and Perils of Medical Enhancement* (New York: Vintage Books, 2004), 12.

36. Oliver Kamn, T. B. Aldrich, I. W. Grote, L. W. Towe, and E. P. Bugbee, "The Active Principles of the Posterior Lobe of the Pituitary Gland," *Journal of the American Chemical Society* 50 (1928): 573–601; on "therapeutic indications," see 599.

37. Ely and Petersen, "Factors Involved," 221.

38. Ibid., 220.

39. Ibid., 221.

40. Newton and Newton, "The Let-Down Reflex in Human Lactation," 698–704.

41. Clair Isbister, "A Clinical Study of the Draught Reflex in Human Lactation," *Archives of Disease in Childhood* 29 (1954): 66–72; Audrey Palm Riker, "Successful Breast Feeding," *American Journal of Nursing* 60, no. 10 (October 1960): 1443–46.

42. M. Hines Roberts [discussant], in Richardson, "Breast Feeding," 867.

43. Augusta Stuart Clay, "Guidance in Maternal and Infant Care Two Months before and after the Birth of the First Born," *Pediatrics* 2, no. 2 (August 1948): 200–206, quotation on 203.

44. Ibid., 202.

45. Judith Walzer Leavitt, *Make Room for Daddy: The Journey from Waiting Room to Birthing Room* (Chapel Hill: University of North Carolina Press, 2009), 34–47.

46. T. Berry Brazelton, "Effect of Maternal Medication on the Neonate and His Behavior," *Journal of Pediatrics* 58 (1961): 513; Reuben E. Kron, Marvin Stein, and Katharine E. Goddard, "Newborn Sucking Behavior Affected by Obstetric Sedation," *Pediatrics* 37, no. 6 (June 1966): 1012–16.

47. Wolf has also commented on the generational divide in medical approaches to infant feeding. See Wolf, *Don't Kill Your Baby*, 187–89.

48. This has been dubbed "the new pediatrics." See Sydney A. Halpern, *American Pediatrics: The Social Dynamics of Professionalism, 1880–1980*

(Berkeley: University of California Press, 1988), 110; and Alexandra Minna Stern and Howard Markel, eds., *The Formative Years: Children's Health in the United States, 1880–2000* (Ann Arbor: University of Michigan Press, 2002), 11–13.

49. Halpern, *American Pediatrics*, 128.

50. Ibid., 130; Richard W. Olmsted, Ruth I. Svibergson, and James A. Kleeman, "The Value of Rooming-In Experience in Pediatric Training," *Pediatrics* 3, no. 5 (May 1949): 617–21; on the under-training of pediatricians on breastfeeding, see 619; Robert W. Deisher, Alfred J. Derby, and Melvin Sturman, "The Practice of Pediatrics—Changing Trends in Pediatric Practice," *Pediatrics* 24, no. 4 (April 1960): 711–16; on the nature of pediatric practice and education, see 711, 713, and 716; Sidney R. Kemberling, "Supporting Breast-Feeding," *Pediatrics* 63, no. 1 (January 1979): 60–63.

51. Hilde Bruch and Donovan J. McCune, "Psychotherapeutic Aspects of Pediatric Practice," *Pediatrics* 2, no. 4 (October 1948): 405–9, quotation on 405. See also Hilde Bruch, "Psychiatric Aspects of Changes in Infant and Child Care," *Pediatrics* 10, no. 5 (November 1952): 575–80. Bruch and McCune held appointments in the Departments of Psychiatry and Pediatrics at Columbia University.

52. See, for example, Randolph K. Byers, "The Pediatrician and the Psychiatrist," *Pediatrics* 30, no. 5 (November 1962): 679–80, see 679.

53. Edwin J. DeCosta, Milton S. Mark, and Ralph A. Reis, "To Nurse or Not to Nurse," *GP* 3, no. 5 (May 1951): 47–51, quotation on 50.

54. Halpern, *American Pediatrics*, 136.

55. Ibid., 133.

56. Ibid., 129.

57. Bruch, "Psychiatric Aspects of Changes in Infant and Child Care," 579.

58. Frederick C. Irving, "Fifty Years of Medical Progress: Medicine as Science: Obstetrics," *New England Journal of Medicine* 244, no. 3 (1951): 91–100, quotation on 98.

59. Ibid.

60. Phillip E. Rothman, "A Note on Demand Feeding," *American Journal of Obstetrics and Gynecology* 65, no. 3 (March 1953): 651–53, quotation on 651.

61. Ibid., 651.

62. Edith B. Jackson provides a thorough literature review of data from the United States and Europe in her article, Edith B. Jackson et al., "Statistical Report on Incidence and Duration of Breast Feeding in Relation to Personal-Social and Hospital Maternity Factors," *Pediatrics* 17, no. 5 (May 1956): 700–715.

63. Herman F. Meyer, "Breast Feeding in the United States: Extent and Possible Trend," *Pediatrics* 22, no. 1, part 1 (July 1958): 116–121, see 117.

64. Claire Rayner, "Feeding the Newborn Infant," *Medical World* 94 (1961): 256–57, quotation on 256.

65. Anderson, "Infant Feeding," 211.

66. Robert L. Jackson, Roy Westerfeld, Margaret A. Flynn, E. Robbins Kimball, and Ray B. Lewis, "Growth of 'Well-born' American Infants Fed Human and Cow's Milk," *Pediatrics* 33, no. 5 (1964): 642–52, and 645 for growth curve comparisons. Infants fed breast milk tend to grow more quickly in the first several months than do formula-fed infants, but then are surpassed in growth by formula-fed babies. In 2010, the U.S. Centers for Disease Control finally recommended that American pediatricians use growth curves based on breastfed infants as the basis for evaluating growth and development for the first two years of life. Up until that point, CDC data was based primarily on measurement data from formula-fed infants, which made it look as though breastfed infants did not grow "as well as" those fed on formula in the long run. See Laurence M. Grummer-Strawn, Chris Reinold, and Nancy F. Krebs, "Use of World Health Organization and CDC Growth Charts for Children Aged 0–59 Months in the United States," *Morbidity and Mortality Weekly Report* (10 September 2010), 1–15, available online at http://www.cdc.gov/mmwr/preview/mmwrhtml/rr5909a1.htm (accessed 18 August 2014); L. J. Filer, Jr., "Commentary: Infant Feeding in the 1970s," *Pediatrics* 47, no. 3 (1971): 489–90, see 490.

67. "Infant Feeding," *Medical Times* 78, no. 11 (November 1950): 501–13, see 501; Lewis A. Barness, "Infant Feeding: Formula Feeding," *Pediatric Clinics of North America* 8 (May 1961): 639–49, see 642.

68. J. R. Ring, A. J. Vignec, and D. J. Donovan, "A Sterile Disposable Nurser System for Infant Feeding," *Hospital Progress* 43 (September 1962): 78–82.

69. L. J. Filer, Jr., "Commentary," 489; on formula morbidity and mortality, see Derrick B. Jelliffe, "Culture, Social Change and Infant Feeding," *American Journal of Nutrition* 10 (January 1962): 19–45, see 35–36; Fischel J. Coodin and Ira W. Gabrielson, "Formula Fatality" *Pediatrics* 47, no. 2 (February 1971): 438–39.

70. James E. Mebs, "New Infant Feeding Procedures Free Formula Room for Other Activities," *Hospital Management* 98 (November 1964): 102–5.

71. "Some Economic and Social Aspects of Infant Feeding," *Nutrition Review* 25, no. 8 (August 1967): 255–56, quotations on 256.

72. Charles D. May, "Determination and Significance of a Dietary Allowance of Protein for Infants . . . *Borden Award Address*," *Pediatrics* 23, no. 2 (February 1959): 384–99.

73. For a discussion of Spock's influence on the field of pediatrics, see Bart Barnes, "Pediatrician Benjamin Spock Dies," *Washington Post* (17 March 1998), A01; and Morris A. Wessel, "How I Got to Be What I Wanted to Be," *Pediatrics* 102, no. 2 (1 August 1998): 384–88.

74. Stevenson, "Adequacy of Artificial Feeding in Infancy," 617.

75. Rayner, "Feeding the Newborn Infant," 256.

76. Smith et al., "Present Day Attitudes towards Breast Feeding," 658; Morrison, "Breast Feeding," 1476; William Oberman and Frederic G. Burke, "Infant

Feeding," *GP* 11, no. 1 (January 1955): 79–86, on emotional immaturity, see 82; see DeCosta, Mark, and Reis, "To Nurse or Not to Nurse," especially 48 for descriptions of mothers who found nursing repulsive.

77. Hans G. Keitel and Norma B. Keitel, "The Clinical Objectives of Infant Feeding," *GP* 24, no. 2 (August 1961): 83–88, on stresses of breastfeeding, see 84; Henry H. Work, "Round Table Discussion: Preventive Psychiatry," *Pediatrics* 10, no. 1 (July 1952): 60–67, and 67 on husbands; Sherman Little, "Round Table Discussion: Preventive Psychiatry," *Pediatrics* 10, no. 1 (July 1952): 68–74, and 71 on father's relationship; DeCosta, Mark, and Reis, "To Nurse or Not to Nurse," and 50 on anxiety and guilt of the mother.

78. Smith et al., "Present Day Attitudes," 658.

79. Little, "Round Table Discussion," 72.

80. S. C. Henn (discussant), in E. Robbins Kimball, "Breast Feeding in Private Practice," *American Journal of Disease of Children* 83, no. 4 (April 1952): 511–13, see 512.

81. Harold Waller, "Some Clinical Aspects of Lactation," *Archives of Disease in Childhood* 22 (December 1947): 193–99, quotation on 198.

82. T. Berry Brazelton, "The Early Mother-Infant Adjustment," *Pediatrics* 32, no. 5 (November 1963): 931–37.

83. Kron, Stein, and Goddard, "Newborn Sucking Behavior Affected by Obstetric Sedation," 1012.

84. Letter to Dr. Spock, c.1957, Folder November 1957, Box 3, Group 1, Benjamin Spock Personal Papers, Special Collections Research Center, Syracuse University, Syracuse, New York (hereafter BSPP). I have purposely removed names and other identifying information from the excerpts and citations of mothers' letters to Spock out of respect for privacy and at the behest of Syracuse University's Special Collections Research Center.

85. Ibid. Despite ample evidence that these freebies can influence a mother's success with breastfeeding, the phenomenon of formula giveaways, for example, continues to be an issue at hospitals around the United States. Kenneth D. Rosenberg, Carissa A. Eastham, Laurin J. Kasehagen, and Alfredo P. Sandoval, "Marketing of Infant Formula through Hospitals: The Impact of Commercial Hospital Discharge Packs on Breastfeeding," *American Journal of Public Health* 98, no. 2 (February 2008): 290–95; Radha Sadacharan, Xena Grossman, Emily Sanchez, and Anne Merewood, "Trends in U.S. Hospital Distribution of Industry-Sponsored Infant Formula Sample Packs," *Pediatrics* 128, no. 4 (26 September 2011): 702–5; Pam Belluck, "Hospitals Ditch Formula Samples to Promote Breastfeeding," *New York Times* (15 October 2012), http://www.nytimes.com/2012/10/16/health/hospitals-ditch-formula-samples-to-promote-breast-feeding.html?pagewanted=all (accessed 18 August 2014).

86. For more on pediatricians' reliance on weight as a health indicator, see Jeffrey P. Brosco, "Weight Charts and Well Child Care: When the Pediatri-

cian Became the Expert in Child Health," in *The Formative Years: Children's Health in the United States, 1880–2000*, ed. Alexandra Minna Stern and Howard Markel (Ann Arbor: University of Michigan Press, 2002), 91–120, see 94.

87. Waller, "Some Clinical Aspects," 197.

88. Leavitt, *Brought to Bed*, 171; Wolf, *Deliver Me from Pain*, 105–35.

89. "Infant Feeding," *Medical Times*, 501–13; "Report of Committee on Mother's Milk Bureaus," *Pediatrics* 1 (1948): 109; Frank H. Richardson, *The Nursing Mother* (New York: David McKay Company, 1953), 93.

90. Letter to J. Kelliher, 13 March 1957, Folder 3, Box 2, BACE Records, SLSC.

91. Morrison, "Breast Feeding," 1481.

92. Ibid.

93. Edith B. Jackson, Louise Wilkin, and Harry Auerbach, "Statistical Report on Incidence and Duration of Breast Feeding in Relation to Personal-Social and Hospital Maternity Factors," *Pediatrics* 17, no. 5 (May 1956): 700–715, quotation on 706.

94. Frederick W. Goodrich and Herbert Thoms, "A Commentary on Natural Childbirth," *Pediatrics* 3, no. 5 (May 1949): 613–16, see 613.

95. Jackson et al., "Statistical Report," 710.

96. Ibid., 711.

97. "Infant Feeding," *Medical Times*, 506.

98. Morrison, "Breast Feeding," 1476, 1478; E. Robbins Kimball, Willard Z. Kerman, Eugene T. McEnery, Gerard N. Krost, and S. C. Henn, "Society Transactions: Breast Feeding in Private Practice," *American Journal of Diseases of Children* 83, no. 4 (April 1952): 511–13, see Dr. S. C. Henn's discussion on "poor nipples" on 512, and see Dr. Kimball's discussion on nipple shields on 513; Richardson, *The Nursing Mother*, 47–51.

99. Harry Bakwin, "Infant Feeding," *American Journal of Clinical Nutrition* 1, no. 5 (July–August 1953): 349–54, see 351.

100. E. Robbins Kimball, "Breast Feeding in Private Practice," *Quarterly Bulletin of Northwestern University Medical School* 25 (1951): 257–62, quotation on 257.

101. Wiessinger, West, and Pitman, *Womanly Art of Breastfeeding* (2010), 62–81. La Leche League's breastfeeding manual contains an entire chapter devoted to this subject, "Latching and Attaching."

102. Breastfeeding advocates who studied the finer details of lactation often advocated manual expression because it correlated with higher breastfeeding success rates. See, for example, E. J. Huenekens, "Breast Feeding," *American Journal of Nursing* 24, no. 9 (June 1924): 751–57, see 754–55; Norman M. MacNeil, "Can We Retrieve Breast-Feeding?" *Pennsylvania Medical Journal* 51, no. 2 (1947): 137–140, see 138; Bakwin, "Infant Feeding," 351; and Muriel H. McClure, "When She Chooses Breast Feeding," *American Journal of Nursing*, 57, no. 8 (August 1957): 1002–5, see 1004.

103. Richardson, "Breast Feeding Comes of Age," 865.

104. "Infant Feeding," *Medical Times*, 503.

105. Morrison, "Breast Feeding," 1477.

106. Kimball, "Breast Feeding in Private Practice," 257.

107. For example, see Oberman and Burke, "Infant Feeding," 79; McClure, "When She Chooses Breast Feeding," 1002; Barness, "Infant Feeding," 641; L. Emmett Holt, Jr., and Selma E. Snyderman, "The Feeding of Premature and Newborn Infants," *Pediatric Clinics of North America* 13, no. 4 (November 1966): 1103–15, see 1112; R. M. Applebaum, "The Modern Management of Successful Breast Feeding," *Pediatric Clinics of North America* 14, no. 1 (February 1970): 203–25, see 203; Nutrition Committee of the Canadian Paediatric Society, "Breast Feeding," 597.

108. "Breast Feeding: Booklet Number Two," Mt. Sinai Hospital of Cleveland, Folder—Medical Reference Files—Breast Feeding, Box 28, Group 1, BSPP; Niles Newton, "Nipple Pain and Nipple Damage," *Journal of Pediatrics* 41, no. 4 (October 1952): 411–23. Newton sought to understand the origins of nipple pain and the best way to manage it in a study of 287 mothers in the rooming-in maternity wards of Philadelphia's Jefferson Hospital. She reported in her 1952 article that mothers who used soap and water or alcohol solutions on their nipples were most likely to experience difficulties with nipple pain and cracking. She also denounced the widespread practice of limiting sucking time (411).

109. Edith B. Jackson, "New Trends in Maternity Care," *American Journal of Nursing* 55, no. 5 (May 1955): 584–87, quotation on 586; Smith et al., "Present Day Attitudes," 656.

110. Emily P. Bacon, "Practical Aspects of Infant Feeding," *Medical Clinics of North America* 36, no. 6 (November 1952): 1555–60, quotation on 1557.

111. Ibid.

112. G. E. Egli, N. S. Egli, and Michael Newton, "The Influence of the Number of Breast Feedings on Milk Production," *Pediatrics* 27, no. 2 (February 1961): 314–17, quotation on 317.

113. Newton and Newton, "Recent Trends," 695.

114. Ibid.

115. Lee Forest Hill, "Infant Feeding: Historical and Current," *Pediatric Clinics of North America* 14, no. 1 (February 1967): 255–68, see 263.

116. Cahill, *Seven Voices, One Dream*, 6–12.

117. Edith B. Jackson to R. G., 26 February 1968, Physicians, 1956–2005, Jackson, Edith B. (1968–1977), LLLIR.

118. "Edith Banfield Jackson, M.D.," (c.1977), Physicians, 1956–2005, Jackson, Edith B. (1968–1977), LLLIR.

119. Ibid.

120. "La Leche League International Third Biennial Convention" (c.1968), Physicians, 1956–2005, Mendelsohn, Robert (1964–1979), LLLIR.

121. Dennis Hollins, "Maverick Physician Calls Medical 'Religion' Harmful," *Buffalo Evening News* (4 February 1980), Physicians, 1956–2005, Mendelsohn, Robert—Correspondence (1964–1979), LLLIR.

122. Letter to Spock, 29 March 1960, Folder March 1960, Box 6, Group 1, BSPP.

123. Hill, "Infant Feeding," 263.

124. Edith B. Jackson to La Leche League, Intl., 26 February 1968, Physicians, 1956–2005, Jackson, Edith B. (1968–1977), LLLIR.

125. Wolf, *Don't Kill Your Baby*, 189–97.

126. Hill, "Infant Feeding," 262.

127. Leavitt, *Make Room for Daddy*, 195–283. As Leavitt shows, some hospitals "continued to be pockets of inequality of access" to more humane delivery practices into the 1980s (284).

Chapter 3

1. Eleanor Lake, "Breast Fed Is Best Fed," *Reader's Digest* (June 1950) [reprint], Member File—Martha Pugacz, LLLIR.

2. On the pull of "natural" mothering in this period, see Grant, *Raising Baby by the Book*, 209–11. On the "emotionalization" of female obligations and roles more generally, see Ruth Schwartz Cowan, "The Industrial Revolution in the Home," in *The Social Shaping of Technology*, ed. Donald MacKenzie and Judy Wajcman (Philadelphia: Open University Press, 1999 [orig. 1985]), 181–201.

3. Maja Bernath, "Bottle for Baby with Love," *Parents' Magazine and Better Homemaking* 35 (May 1960): 38, 126–27.

4. B. F. Skinner, "Baby in a Box," *Ladies' Home Journal* 62 (October 1945), 30–31, 135–36, 138.

5. B. F. Skinner "The First Baby Tender," *Behaviorology Today* 7, no. 1 (2004): 2. Originally published in Stephen F. Ledoux and Carl D. Cheney, eds., *Grandpa Fred's Baby Tender, or Why and How We Built Our Aircrib* (Canton, NY: ABCs, 1987).

6. Jill G. Morawski, "Educating the Emotions: Academic Psychology, Textbooks, and the Psychology Industry, 1890–1940," in *Inventing the Psychological: Toward a Cultural History of Emotional Life in America*, ed. Joel Pfister and Nancy Schnog (New Haven, CT: Yale University Press, 1997); Jill G. Morawski and Gail A. Hornstein "Quandary of the Quacks: The Struggle for Expert Knowledge in American Psychology, 1890–1940," in *Estate of Social Knowledge*, ed. JoAnne Brown and David K. van Keuren (Baltimore: Johns Hopkins University Press, 1991), 106–33.

7. Cowan, *More Work for Mother.*

8. For more on Skinner and his public image, see Alexandra Rutherford,

"Radical Behaviorism and Psychology's Public: B. F. Skinner in the Popular Press, 1934–1990," *History of Psychology* 3, no. 4 (2000): 371–95; and Alexandra Rutherford "B. F. Skinner's Technology of Behavior in American Life: From Consumer Culture to Counterculture," *Journal of the History of the Behavioral Sciences* 39 (2003): 1–23.

9. See, for example, *Infant Care*, U.S. Children's Bureau Publication No. 8 (Federal Security Agency, Social Security Administration, 1945); Niles Newton, *Family Book of Child Care* (New York: Harper & Row, 1957); *Parents' Magazine's Baby Care Manual* (New York: Parents' Institute, 1941); Benjamin Spock, *The Common Sense Book of Baby and Child Care* (New York: Sloane and Pierce, 1945). Businesses also helped disseminate infant feeding advice through the mail. For example: "Baby's Record Book," Metropolitan Life Insurance Company (New York, c.1945); "Your Child—The First Year of Life," The Prudential Insurance Company of America (New York: Greystone Press, 1954); "Happy Mealtimes for Your Baby," Beech-Nut Baby Foods (Canajoharie, NY: Beech-Nut Life Savers, 1960).

10. *Readers' Guide Retrospective: 1890–1982 (H .W. Wilson)*, EBSCOhost (accessed 8 December 2012). The author conducted a survey of over 160 articles through the H. W. Wilson Readers' Guide Retrospective between 1940 and 1963 using search terms "infant or baby and feeding," "infant and milk," "breast or mother and milk," and "infant and bottle or breast." The author also searched key publications, including *Hygeia*, *Better Homes and Gardens*, *Today's Health*, *Good Housekeeping*, *Ladies' Home Journal*, *Redbook*, *McCall's*, *Parents' Magazine*, *Science News Letter*, and *Reader's Digest*. See, for example, G. D. Schultz, "Nurse Your Baby? Sure You Can!" *Better Homes and Gardens* 22 (July 1944), 40; Herman N. Bundesen, "Best-Fed Are Breast-Fed," *Ladies' Home Journal* 61 (July 1944), 128; "The Doctor Talks about Breast-Feeding," *McCall's* 84 (May 1957), 4; Mabel C. Woodard, "Breast or Bottle for a Baby?" *Parents' Magazine* 20 (August 1945), 20 and following; Jean Y. Harshberger, "Getting Ready to Nurse," *Parents' Magazine* 24 (November 1949), 37 and following.

11. Bruce Gould to Spock, 28 November 1956, Folder November 1956, Box 3, Group 1, BSPP.

12. "Faulty Sex Relations Stem from Bottle-Feeding Baby," *Science News Letter* 56 (22 October 1949); Mead, *Male and Female*, 268–69.

13. Benjamin Spock to Edith B. Jackson, 16 September 1955, Folder 140, Box 7, Edith Banfield Jackson Papers, SLSC.

14. Katherine Clifford, "More TLC If You Can't Breast-Feed Your Baby," *Parents' Magazine* 27 (April 1952), 42–43, 88.

15. Letter to Spock, 26 May 1959, Folder May 1959, Box 5, Group 1, BSPP.

16. Newton and Newton, "Recent Trends in Breast Feeding," 691–99; Stevenson, "Adequacy of Artificial Feeding in Infancy," 621 and 623.

17. Gladys Denny Schultz, "Why Can't Our Mother's Breast-Feed?" *Ladies' Home Journal* 67 (December 1950), 42–43, 190–92, 192.

18. Howard J. Morrison, "Breast Feeding: A Doctor Makes an Outspoken Plea for the Original Method of Feeding Babies," *Today's Health* 29 (August 1951): 40.

19. Letter to Spock, 30 July 1954, Folder September 1954, Box 2, Group 1, BSPP.

20. "Questions & Answers: Breast Feeding," *Today's Health* 24 (January 1946): 79; Jessica Martucci, "Maternal Expectations: New Mothers, Nurses, and Breastfeeding," *Nursing History Review* 20 (2012): 72–102, 81.

21. Letter to William Cochran, 25 January 1963, Folder 10, Box 6, BACE records, SLSC.

22. J. Irwin Foss, "Let's Break the Taboo against Breast Feeding," *Hygeia* 27 (January 1949): 20–21, 69.

23. Woodard, "Breast or Bottle for Baby?" 20–21, 78, 80, 82, 84.

24. Mary Margaret Kern, "Two Breast Nursing," *Parents' Magazine* 21 (November 1946), 23, 154, 169–172, quotation on 23.

25. Ibid., 23.

26. Foss, "Let's Break the Taboo," 69.

27. Woodard, "Breast or Bottle?" 21.

28. Letter to Spock, 30 July 1954, Folder September 1954, Box 2, Group 1, BSPP.

29. Letter to Jackson, 14 September 1953, Folder 50, Box 1, Edith Banfield Jackson Papers, SLSC.

30. Ibid.

31. Letter to Spock, 23 June 1956, Folder June 1956, Box 3, Group 1, BSPP.

32. Letter to Edith Jackson, 18 June 1954, Folder 50, Box 1, Edith Banfield Jackson Papers, SLSC.

33. "It's New Again [Breast Feeding]," *Hygeia* 26 (August 1948), 587.

34. See, for example, Holt, *Care and Feeding of Children*, 42; Watson, *Psychological Care*, 3–4; Aldrich, *Babies Are Human Beings*, 70–71; and Ribble, *Rights of Infants*, 33, 105.

35. Spock, *Common Sense Book*, 33; Vanta Company, *Your Baby and You: Vanta's Book for Mothers and Mothers-to-Be*, 7th ed. (Newton, MA: Vanta Co., 1949), 30, author's collection; Carnation Company, *You and Your Contented Baby* (Los Angeles: Carnation Co., 1956), 28, author's collection; Newton, *Family Book of Child Care*, 80–82.

36. On 27 March 1958, Dr. Herbert Ratner mediated a formal discussion with the seven League founders: Mary Ann Cahill, Edwina Froehlich, Mary Ann Kerwin, Marian Tompson, Betty Wagner, and Mary White. Meeting minutes indicate the founders discussed making women "better" mothers through teaching

breastfeeding. "La Leche League Dialogue: An Historic Document" (reprint, 1981), LLLIR.

37. Herbert Ratner authored the foreword to the League's *The Womanly Art of Breastfeeding*, which remained throughout several editions into the 1980s. Ratner described America as a "sick society," which embraced divorce and unnecessary technology, among other things. Letter to La Leche League, August 1980, Folder 1, Box 7—Board Correspondence, LLLIR.

38. Leavitt, *Make Room for Daddy*, 239–43. Leavitt's argument about fathers entering the birthing room highlights men's changing domestic roles in the later twentieth century.

39. Eleanor S. Duncan and Dorothy Whipple, *The New Parents' Magazine Baby Care Book* (New York: Parents' Magazine Enterprises, 1969), 7.

40. Jay Mechling, "Advice to Historians on Advice to Mothers," *Journal of Social History* 9, no. 1 (1975): 44–63.

41. May, *Homeward Bound*, 9–26.

42. I purposely utilize the terminology of "spheres" here to suggest that breastfeeding supporters in this period actively staked a claim for maternal authority within the contested space of the postwar home and family. Kim Warren, "Separate Spheres: Analytical Persistence in United States Women's History," *History Compass* 4 (2006): 1–16; Linda Kerber, "Separate Spheres, Female Worlds, Woman's Place: The Rhetoric of Women's History," *Journal of American History* 75, no. 1 (June 1988): 9–39.

43. *Womanly Art* (1963), 114–15.

44. Alice Von Briesen, "Eight Reasons Why," *Parent's Magazine* 24 (April 1949), 30, 125.

45. Martha Pugacz to Edwina Froehlich, 7 December 1958, Members Files—Martha Pugacz, LLLIR.

46. Newton, *Maternal Emotions*, x–2.

47. Samuel Wishik, Nancy Dingman Watson, and Adeline Bullock, "Breast, Bottle, or Both for Your Baby," *Parents' Magazine and Family Home Guide* 31 (September 1956), 131.

48. Letter to Spock, 20 June 1956, Folder June 1956, Box 3, Group 1, BSPP.

49. Russell C. Smart, "In Defense of Bottles for Babies," *Parents' Magazine* 25 (April 1950), 35.

50. Ibid.

51. Margaret O'Keefe, "We're Twenty Years Behind in Breast-Feeding," *Ladies' Home Journal* 79 (November 1962), 134.

52. Robert B. Westbrook, "'I want a girl, just like the girl that married Harry James': American Women and the Problem of Political Obligation in World War II," *American Quarterly* 42, no. 4 (December 1990): 587–614.

53. Ruth and Edward Brecher, "What They've Been Learning about Breast Feeding," *Ladies' Home Journal* 72 (1955), 26–28.

54. Brecher, *Sex Researchers*, 175.

55. For example, see Olga Bonke-Booher, "20 Questions and Answers on Baby Feeding," *Parents' Magazine and Family Home Guide* 29 (May 1954), 63, 94–95; Foss, "Let's Break the Taboo," 69; Benjamin Spock, "Getting Started on Breast Feeding," *Ladies' Home Journal*, 74 (April 1957), 20, 23, 25–26.

56. Westbrook, "'I want a girl,'" 587–614.

57. Brumberg, *Body Project*, 109–18.

58. Jay Smith, "The Big Bosom Battle," *Playboy* 2, no. 9 (September 1955), 23.

59. Jane Farrell-Beck and Colleen Gau, *Uplift: The Bra in America* (Philadelphia: University of Pennsylvania Press, 2002), 121. Unfortunately I was unable to obtain permission to reprint this image from *Playboy*.

60. Blum, *At the Breast*, 38.

61. Ibid., 40.

62. Marilyn Yalom, *A History of the Breast* (New York: Ballantine Books, 1998) 5, 132.

63. *Womanly Art* (1963), 19.

64. Spock, *Common Sense Book*, 32.

65. Stephanie Knaak, "Breast-Feeding, Bottle-Feeding and Dr. Spock: The Shifting Context of Choice," *Canadian Review of Sociology and Anthropology* 42, no. 2 (2005): 197–216.

66. Betsy Marvin McKinney, "The Sexual Aspect of Breast Feeding," *Child-Family Digest* 13, no. 4 (December 1955): 45–47.

67. Kinsey et al., *Sexual Behavior in the Human Female* (Bloomington: Indiana University Press, 1998 [orig. 1953]), 586.

68. Ibid., 253.

69. May, *Homeward Bound*, 114–24.

70. Ilse S. Wolff, "Mothers' Views on Breast Feeding," *Nursing Outlook* 1, no. 3 (March 1953): 145–48.

71. Letter to Spock, 12 December 1959, Folder December 1959, Box 5, Group 1, BSPP.

72. Letter to Spock, 29 March 1960, Folder March 1960, Box 6, Group 1, BSPP.

73. Letter to Spock, c.1957, Folder February 1957, Box 3, Group 1, BSPP.

74. Harshberger, "Getting Ready to Nurse [Right Care of Breasts and Nipples]," 37, 86–88, quotation on 37.

75. "Your Baby's Care" [pamphlet] (St. Paul, MN: Carleton J. West Publications, 1959), 20, author's collection.

76. Newton, *Maternal Emotions*.

77. Letter to Spock, 3 February 1956, Folder February 1956, Box 3, Group 1, BSPP.

78. Letter to Spock, 18 March 1957, Folder March 1957, Box 3, Group 1, BSPP.

79. Donna Penn, "The Meanings of Lesbianism in Post-War America," *Gender and History* 3, no. 2 (1991): 190–203.

80. Letter to Spock, 29 March 1960, Folder March 1960, Box 6, Group 1, BSPP.

81. *Womanly Art* (1963), 13.

82. Ibid.

83. Letter to Medical Advisory Board, 15 August 1970, Folder 2, Box 2, Physicians, 1956–2005, LLLIR.

84. Eleanor Lake, "Breast Fed Is Best Fed," *Reader's Digest* (June 1950) [reprint], Member File—Martha Pugacz, LLLIR.

85. Plant, *Mom*, 2–6.

86. Letter to Martha Pugacz, 12 March 1959, Members Files—Martha Pugacz, LLLIR.

Chapter 4

1. McClure, "When She Chooses Breast Feeding," 1002.

2. Lorraine Dyal and Julia Kahrl, "When Mothers Breast-Feed," *American Journal of Nursing* 67, no. 12 (December 1967): 2555.

3. For example, see Bruch and McCune, "Psychotherapeutic Aspects of Pediatric Practice," 405–6; Herman F. Meyer, "Breast Feeding in the United States," 116–21; Niles Newton and Michael Newton, "Relation of the Let-Down Reflex to the Ability to Breastfeed," *Pediatrics* 5, no. 4 (April 1950): 726–33; and McClure, "When She Chooses Breast Feeding," 1002–5.

4. Wolf, "What Feminists Can Do," 397–424.

5. Daniel Chambliss, *Beyond Caring: Hospitals, Nurses, and the Social Organization of Ethics* (Chicago: University of Chicago Press, 1996).

6. See, for example, Margarete Sandelowski, *Devices and Desires: Gender, Technology, and American Nursing* (Chapel Hill: University of North Carolina Press, 2000); and Judith Walzer Leavitt, "'Strange Young Women on Errands': Obstetric Nursing between Two Worlds," *Nursing History Review* 6 (1998): 3–24. In her work on labor and delivery nurses, Leavitt expands the scope of gender analysis in the hospital to include the ways in which gender influenced same-sex interactions between female patients and female nurses from the 1930s through the 1950s, particularly in the context of an experience as highly gendered as the birth and feeding of an infant. Her work suggests that women looked to nurses during their hospital stays for support, as gowned versions of their female friends and family during a distinctly female experience. In addition to demonstrating their desire for a scientifically modern birth under the care of a physician, Leavitt has shown that mothers sought familiarity and female comfort from nurses during their parturition. "Despite all their longings for the benefits of

science," she writes, "they did not—could not—leave behind these other expectations of how women should behave toward one another, especially at such a critical time as labor and delivery" (17). Leavitt's idea that nurses and mothers brought differing expectations about one another to the hospital maternity ward based on their shared gender serves as a basis for my work on nurses and breast-feeding mothers.

7. Mary Thomas, "Grantly Dick-Read and Natural Childbirth: A Turning Point in the History of Childbirth," in *Post-War Mothers: Childbirth Letters to Grantly Dick-Read*, ed. Mary Thomas (Rochester, NY: University of Rochester Press, 1997), 6. See also Leavitt, *Brought to Bed*, 171.

8. Starr, *Social Transformation of American Medicine*, 348; Rosemary Stevens, *In Sickness and in Wealth: American Hospitals in the Twentieth Century* (New York: Basic Books, 1989), 216–19; "Hill-Burton, 1946," *Hospitals and Health Networks* 81, no. 3 (March 2007): 1.

9. Stevens, *In Sickness*, 227.

10. Leavitt, *Make Room for Daddy*, 53.

11. Barbara Melosh, *"The Physician's Hand": Work Culture and Conflict in American Nursing* (Philadelphia: Temple University Press, 1982), 195. The rise of the dynamic of nurse-mother meets mother in the hospital became more and more likely in the decades following World War II. By 1951, for example, Melosh observed that 47 percent of all active nurses were married, a dramatic increase from earlier in the century. This number only increased during the 1950s and 1960s. One study showed, for example, that by 1958, 55 percent of active nurses were married, a rate that was much higher than for the general female worker population. See also Ann Marie Walsh Brennan, *Fifty Nurses over Fifty: Prominent Themes within the Work Histories of Persistently Employed Women* (PhD diss., University of Pennsylvania, 1997), 86.

12. Evidence for the interactions of mothers and their nurses is drawn from correspondence files in Benjamin Spock Personal Papers at Syracuse University Special Collections, La Leche League International Collection at DePaul University Special Collections, and the Boston Women's Health Book Collective, Edith B. Jackson, and Boston Association for Childbirth Education collections at the Schlesinger Library, Radcliffe Institute, Harvard University. Additional materials are drawn from the collections of the Barbara Bates Center for the Study of the History of Nursing at the University of Pennsylvania, as well as from published accounts in the *American Journal of Nursing* and women's popular magazines from the era.

13. *Womanly Art* (1963), 20.

14. Gladys Denny Shultz, "Why Can't Our Mothers Breast-Feed?" *Ladies' Home Journal* 67 (December 1950), 42–43, 191–93.

15. Audrey Palm Riker, "What about Breast and Bottle-Feeding?" *Parents' Magazine and Better Homemaking* 36 (July 1961), 32–33, 96–98.

16. *Womanly Art* (1963), 53–54.

17. Letitia Lyon Sage, "The Battle of the Bottle: An Inexperienced Mother Needs Help and Reassurance from the Nurse If Breast Feeding Is to Be Established," *American Journal of Nursing* 47, no. 6 (1947): 395.

18. Ibid.

19. Erin Anderson and Elizabeth Geden, "Nurse's Knowledge of Breastfeeding," *Journal of Obstetric, Gynecologic, and Neonatal Nursing* 20, no. 1 (1991): 58–62. On more recent discussions of medical knowledge about breastfeeding, see Wolf, "Low Breastfeeding Rates and Public Health in the United States," 2000–2010; and Wolf, "What Feminists Can Do," 397–424.

20. Riker, "Successful Breast Feeding," 1444; Sister Joseph Sarto and Margaret O'Keefe, "Breast Feeding," *American Journal of Nursing* 63, no. 12 (December 1963): 58–60.

21. Riker, "Successful Breast Feeding," 1444; Sarto and O'Keefe, "Breast Feeding," 58–60.

22. Lorraine Weszely, "Breast-Feeding Help Needed," *American Journal of Nursing* 68, no. 7 (1968): 1441–46, quotation on 1441.

23. *Womanly Art* (1963), 53–54.

24. Ibid.

25. Samuel Lubin, "The Routine Use of Stilbestrol for Engorgement and Lactation in Nonnursing Mothers," *American Journal of Obstetrics and Gynecology* 51 (February 1946): 225–29; Colin Hodge, "Suppression of Lactation by Stilboestrol," *Lancet* 290, no. 7510 (August 1967): 286–7; Daniel J. Schwartz et al., "A Clinical Study of Lactation Suppression," *Obstetrics and Gynecology* 42, no. 4 (October 1973): 599–606.

26. Sergio C. Stone and Richard P. Dickey, "Management of Nursing and Nonnursing Mothers," *Clinical Obstetrics and Gynecology* 18, no. 2 (June 1975): 139–48, quotation on 144.

27. Birth report [1970], Folder 2, Box 13, BACE Records, SLSC.

28. U.S. Department of Health and Human Services, Public Health Service, National Toxicology Program, "Diethylstilbestrol CAS No. 56–53–1," in *Report on Carcinogens, Eleventh Edition* (2005). Also known as DES, or diethylstilbestrol, the synthetic hormone compound was first put into use in 1941 and became heavily relied upon for the prevention of miscarriages, to induce puberty in girls, and to suppress postpartum lactation. In 1978 the FDA recalled its support of the drug for the suppression of lactation, and in 1980 the drug was officially recognized as a carcinogen.

29. *Womanly Art* (1963), 55.

30. Leavitt, *Make Room for Daddy*, 239–43. Leavitt observes that by the mid-1970s significant progress had been made, for example, in allowing fathers to attend the births of their children in the hospital. She notes, however, that there

remained "significant regional variation," a circumstance mirrored by hospital policies toward breastfeeding as well during this decade.

31. Shultz, "Why Can't Our Mothers Breast-Feed?" 43.

32. Clifford, "More TLC If You Can't Breast-Feed Your Baby," 42.

33. Interview with Barbara (1 October 2008), author's files.

34. Evelyn Kanter, "I Chose to Nurse My Baby," *Parents' Magazine and Better Family Living* 47 (June 1972), 32, 68–9, quote on 32.

35. M. S. S., "Better Babies" *American Journal of Nursing* 47, no. 9 (1947): 638.

36. Marian Wenrich et al., "The Family Care Study," *American Journal of Nursing* 48, no. 3 (1948): 179–84, 180.

37. Ibid.

38. Ibid.

39. Weszely, "Breast-Feeding Help Needed," 1441.

40. Clifford, "More TLC If You Can't Breast-Feed Your Baby," 68.

41. Jackson, "New Trends in Maternity Care," 584.

42. Leavitt, *Brought to Bed*, 12.

43. "Maternal and Newborn Care Conference for Nurses," 21–31 March 1950, Folder 9, Box 4, Edith Banfield Jackson Papers (hereafter EBJP), SLSC.

44. B. V. to Dr. Milton Senn, 10 January 1952, Folder 16, Box 2, EBJP, SLSC.

45. Jackson, "Should Mother and Baby Room Together?" 17.

46. Shirley E. Lundgren, "A Trial Rooming-In Plan," *American Journal of Nursing* 47, no. 8 (1947): 547–48, quote on 548.

47. Ibid.

48. Ibid.

49. Apple, *Perfect Motherhood*, 116.

50. Lundgren, "A Trial Rooming-In Plan," 547.

51. Mary Helen Anderson et al., "There Are Two Sides," *American Journal of Nursing* 47, no. 12 (1947): 806.

52. *Womanly Art* (1963), 151.

53. Margaret O'Keefe, "Advice from a Nurse-Mother," *American Journal of Nursing* 63, no. 12 (1963): 64; Elizabeth Peck and Ruth Carney, "Guidance Programs for New Mothers," *American Journal of Nursing* 51, no. 3 (1951): 186.

54. Betty Ann Countryman, "Hospital Care of the Breast-Fed Newborn," *American Journal of Nursing* 71, no. 12 (1971): 2365–67.

55. Sarto and O'Keefe, "Breast Feeding," 58.

56. Dyal and Kahrl, "When Mothers Breast-Feed," 2555.

57. For more on the role of nurses as leaders in maternal health reform, see Janice Templeton Gay et al., "Reva Rubin Revisited," *Journal of Obstetric, Gynecologic, and Neonatal Nursing* 17, no. 6 (1988): 394–99. See also Rubin, "Basic Maternal Behaviors," *Nursing Outlook* 9 (1961): 683–86; Rubin, "Puerperal

Change," *Nursing Outlook* 9 (1961): 753–55; and Rubin, "Maternal Touch," *Nursing Outlook* 11 (1963): 818–31.

58. Josephine Iorio, "Breast-Feeding Mothers Help Each Other," *American Journal of Nursing* 64, no. 10 (October 1964): 119.

59. Ibid.

60. Ibid.

61. "How the Maternity Nurse Can Help the Breastfeeding Mother," no. 118 (Franklin Park, IL: La Leche League International, 1972), 6.

62. Ibid., 5.

63. Ibid., 10.

64. Ibid., 22.

65. Ibid.

66. Dyal and Kahrl, "When Mothers Breast-Feed," 2555.

67. Evelyn Staus et al., "Breast vs. Bottle," *American Journal of Nursing* 70, no. 10 (1970): 2085–98, quotation on 2097.

68. Sarto and O'Keefe, "Breast Feeding," 60.

69. "Position Statement on the Nurse's Role in Influencing Infant Feeding Practices," 25 June 1983, Boston Women's Health Book Collective Papers (BWHBC), Folder 11, Box 38, SLSC.

70. Carla B. Patton et al., "Nurses' Attitudes and Behaviors That Promote Breastfeeding," *Journal of Human Lactation* 12, no. 2 (1996): 111–15.

71. "How the Maternity Nurse," 5.

72. Ruth E. Owen and Lucille G. Denman, "Experiences in Childbirth," *American Journal of Nursing* 51, no. 1 (1951): 26–27, quotation on 26.

73. O'Keefe, "We're Twenty Years Behind in Breast-Feeding," 70.

74. O'Keefe, "Advice from a Nurse-Mother," 61.

75. Ibid., 62.

76. Ibid.

77. Weszely, "Breast-Feeding Help Needed," 1446.

78. Jean Cotterman, "How to Breast-Feed without the Doctor's Help," *Marriage: The Magazine of Catholic Family Living* 48, no. 2 (February 1966): 13–17, quotation on 13.

79. Barbara Teitelman et al., "Views on Breast Feeding," *American Journal of Nursing* 70, no. 9 (September 1970): 1876–82, quotation on 1876.

80. Ibid.

81. Birth report, 1971, BACE, Folder 4, Box 13, SLSC.

82. Kathleen Knafl, "Conflicting Perspectives on Breast Feeding," *American Journal of Nursing* 74, no. 10 (1974), 1848–51, quotation on 1849.

83. Ibid.

84. Susan Aberman and Karin T. Kirchhoff, "Infant-Feeding Practices: Mothers' Decision Making," *Journal of Obstetric, Gynecologic, and Neonatal Nursing* 14, no. 5 (1985): 394–98.

Chapter 5

1. Interview with Patty (2 February 2009), author's files.

2. Ibid.

3. Ibid.

4. Alan S. Ryan et al., "Recent Declines in Breast-Feeding in the United States, 1984–1989," *Pediatrics* 88, no. 4 (October 1991): 719–27, see 722.

5. "Nursing Mothers" groups grew out of local childbirth education organizations, most of which were affiliated with the International Childbirth Education Association. The one Patty belonged to was likely affiliated with the Nursing Mothers Committee of the Childbirth Education Association of Greater Philadelphia. "Nursing Mothers' Council," BACE, Folder 9, Box 6, SLSC; Interview with Tammy (11 February 2009), author's files.

6. Interview with Patty (2 February 2009), author's files.

7. Lauri Umansky, *Motherhood Reconceived: Feminism and the Legacy of the 1960s* (New York: New York University Press, 1996), 57–63.

8. Wendy Kline has discussed a similar issue in the case of natural childbirth and second-wave feminism. See Kline, *Bodies of Knowledge*, 127–55.

9. Carolyn Merchant, *The Death of Nature: Women, Ecology and the Scientific Revolution* (San Francisco, CA: HarperCollins, 1989 [orig. 1980]), 20–23, 288.

10. E. Melanie DuPuis, *Nature's Perfect Food: How Milk Became America's Drink* (New York: New York University Press, 2002), 94.

11. Ibid., 104.

12. Janet Kushmaul, "Ecology and Breastfeeding," *La Leche League News* 13, no. 6 (November–December 1971): 11, La Leche League Archives, DePaul University Special Collections.

13. Eleanor Agnew, *Back from the Land: How Young Americans Went to Nature in the 1970s, and Why They Came Back* (Chicago: Ivan R. Dee, 2004), 5.

14. Jeffrey Jacob, *New Pioneers: The Back to the Land Movement and the Search for Sustainability* (State College: Pennsylvania State University Press, 1998), 23.

15. J. Rundo, "Fall-Out Caesium-137 in Breast- and Bottle-Fed Infants," *Health Physics* 18, no. 4 (April 1970): 437–38; B. E. Godfrey and J. Vennart, "Measurements of Caesium-137 in Human Beings in 1958–67," *Nature* 218 (London, 1968): 741–46; A. R. Wilson and F. W. Spiers, "Fallout Caesium-137 and Potassium in Newborn Infants," *Nature* 215 (London, 1967): 470–74; A. Aarkrog, "Caesium-137 from Fall-Out in Human Milk," *Nature* 197 (London, 1963): 667–68.

16. Daniel Gross, *Our Roots Grow Deep: The Story of Rodale* (Emmaus, PA: Rodale Press, 2008), 66.

17. Ibid., 68.

18. Letter from Niles Newton, 19 February 1968, Folder—Niles Newton,

Box 80, Physicians Files, LLLIR. Newton's letter includes a reference to and clipping of an article published in *Rodale's Health Bulletin* 6, no. 6 (10 February 1968). The files of Betty Wagner in the LLLIR also contain clippings from Rodale Press publications.

19. For more on the social demographics of these movements, see Ethel Barol Taylor, *We Made a Difference: My Personal Journey with Women Strike for Peace* (Philadelphia: Camino Books, 1998); Vera Norwood, *Made from This Earth: American Women and Nature* (Chapel Hill: University of North Carolina Press, 1993), 143–71; Amy Swerdlow, *Women Strike for Peace: Traditional Motherhood and Radical Politics in the 1960s* (Chicago: University of Chicago Press, 1993); and Cahill, *Seven Voices, One Dream*. A demographic study of the group Women Strike for Peace (WSP) was conducted in 1963, which revealed the vast majority of those involved had obtained some college, and a significant fraction had advanced degrees (21 percent). Nearly two-thirds of WSP's members were between the ages of twenty-five and forty-four, were not employed outside of the home, and nearly half had one to four children under the age of eighteen at the time of the survey. Perhaps the most significant and most uniform statistic to characterize the WSP women and the women throughout these movements was their marriage to a "professional" (70 percent). The author describes her subjects as "happy, non-conformist middle class families united in their goals in a conformist sector of society!" (6). Elise Boulding, "Who Are These Women? A Progress Report on a Study of Women Strike for Peace," [draft] c.1963, Women Strike for Peace Records (hereafter WSP), Box 2, Series A.1, Swarthmore College Peace Collection, Swarthmore, Pennsylvania.

20. Thomas Wellock, *Preserving the Nation: The Conservation and Environmental Movements 1870–2000* (Wheeling, IL: Harlan Davidson, 2007), 166–67.

21. Ibid., 160.

22. "History of SANE, 1957–1963," National Committee for a Sane Nuclear Policy Records (hereafter SANE), Box 1, Swarthmore College Peace Collection.

23. "The Purpose and Program of the National Committee for a Sane Nuclear Policy," c. 1957, Box 2, SANE Records, Swarthmore College Peace Collection.

24. "No Contamination without Representation . . ." *New York Herald Tribune* (24 March 1958) [clipping], Box 12, SANE Records, Swarthmore College Peace Collection.

25. "Dr. Spock Is Worried," *New York Times* (16 April 1962) [reprint], SANE Records; "*Your* Child's Teeth Contain Strontium-90," *New York Times* (7 April 1963) [reprint], SANE Records, Swarthmore College Peace Collection.

26. The degree to which these messages impacted women has been commented on in the literature on environmental history. In Vera Norwood's *Made from This Earth*, the author observes that ecology at midcentury was a "science whose central metaphor for describing nature [was] as a household or home. . . .

In this way, lessons about the web of life learned in scientific ecology meshed with women's knowledge of the web of life they nurtured in their homes" (151).

27. *Womanly Art* (1963), 7.

28. "Strontium-90" Pamphlet, Peninsula Women for Peace (1962), Box 5, Series A.2, WSP Records, Swarthmore College Peace Collection. The group Women Strike for Peace was formed by women from SANE in 1961 in reaction to SANE's increasingly anticommunist policies and its unwillingness to pursue women's issues more strongly. See Swerdlow, *Women Strike for Peace*, 56–57.

29. Jeanne S. Bagby, Committee on Radiation (August 1962), "Suggestions for Reducing Intake of Radioactivity From Foods," WSP Records, Box 5, Series A.2, Swarthmore College Peace Collection.

30. H. S. Telford and J. E. Guthrie, "Transmission of the Toxicity of DDT through the Milk of White Rats and Goats," *Science* 102, no. 2660 (21 December 1945): 647; Edward P. Laug et al., "Occurrence of DDT in Human Fat and Milk," *Archives of Industrial Hygiene* 3 (March 1951): 245–46.

31. Laug et al., "Occurrence of DDT" 245–46. Laug worked for the FDA and was among the first researchers to demonstrate the presence of DDT in the food chain and its presence in human bodies. Transcript, Bert J. Vos, O. Garth Fitzhugh, Edwin P. Laug, Geoffrey Woodard, James Harvey Young, Wallace F. Janssen, Oral History Interview (20 June 1980) by Robert G. Portar and Fred L. Lofsvold, U.S. Food and Drug Administration oral history collection, 1968–2012, OH 81, History of Medicine Division, National Library of Medicine.

32. Carson, *Silent Spring*, 23.

33. C. F. Wurster, Jr., D. H. Wurster, W. N. Strickland, "Bird Mortality after Spraying for Dutch Elm Disease with DDT," *Science* 148 (1965): 90–91; Elena J. Conis, "Debating the Health Effects of DDT: Thomas Jukes, Charles Wurster and the Fate of an Environmental Pollutant," *Public Health Reports* 125, no. 2 (March–April 2010): 337–42.

34. "DDT," *Leaven* 5, no. 6 (November–December 1969): 21. Internal correspondence corroborates the degree to which women were concerned about DDT in breast milk in this period. In a letter dated 24 September 1969 from League secretary Edwina Froehlich to Dr. Herbert Ratner, director of the Department of Public Health in Oak Park, Illinois, Froehlich states, "We are still getting inquiries from mothers, and newspapers are still reporting the dangers of breast milk because of DDT. Everybody seems to think *we* should DO something. Beats me!" Folder—Herbert Ratner, Box 81, Physicians Files, LLLIR.

35. In the July–August 1970 issue of *La Leche League News*, president Marian Tompson wrote, "We do not believe it is within the scope of LLL to take a specific position against its [DDT's] use except as it might affect the nursing mother and baby." However, she did urge mothers not to "stand idly by" and encouraged newsletter readers to write to their elected officials—as individuals.

36. Lana Auriemmo, "Earth Day Was La Leche Speak-Out Day," *LLL News* 12, no. 5 (September–October 1970): 70–71.

37. Paul Shinoff, "Is There a Dangerous Herbicide in Mother's Milk?" *New Pittsburgh Courier* (24 December 1977), 13; "Industrial Chemicals Found in Milk of Nursing Mothers," *Los Angeles Times* (21 August 1976), 6; "Toxic Chemical PBB Found in Breast Milk," *New York Times* (13 October 1976); Staff, "Traces of Kepone Found in Breast Milk in Southern States, EPA says findings may be linked to use of Mirex as bait to kill fire ants," *Wall Street Journal* (27 February 1976), 16.

38. "Strontium 90 in Baby Teeth," *La Leche League News* 7, no. 2 (March–April 1965): 11.

39. "Environmental Contaminants in Mother's Milk," Information Sheet no. 78-a, slightly revised (April 1975), Folder 4, Box 9, Publications, LLLIR.

40. Glenda Daniel, "Mom Finds Her Gift to Baby Is Tainted by Chemicals," *Chicago Tribune* (28 June 1977), A2.

41. Gregory J. White, "Breast-Feeding Benefits," *Chicago Tribune* (7 July 1977), B2.

42. Ibid.

43. Glenda Daniel, "Contaminated Milk," *Chicago Tribune* (23 July 1977), S8.

44. Alan S. Ryan, Zhou Wenjun, and Andrew Acosta, "Breastfeeding Continues to Increase into the New Millennium," *Pediatrics* 110, no. 6 (December 2002): 1103–9. The authors compiled data from over three decades from the Ross Laboratories Mothers Survey, which began collecting data on infant feeding in 1954. In 1971, the Mothers Survey reported 24.7 percent of infants received "any breastfeeding." By 1980, that figure had risen to over 50 percent.

45. It became common for researchers and the media, in general, to portray breast milk as more dangerous than other kinds of infant feeding. For example, in a 1970 report in the journal *Nature* on the harmful effects of DDT on the "nursing neonate," the authors stated that "it has been found that cows excrete less DDT in lactational fluids than human females." Mostafa S. Fahim, Robert Bennet, and David G. Hall, "Effect of DDT on the Nursing Neonate," *Nature* 228 (19 December 1970): 1222–23.

46. Agnew, *Back from the Land*, 154.

47. Ibid.

48. Ibid., 199–200.

49. Samuel J. Fomon, "Infant Feeding in the Twentieth Century: Formula and Beikost," *American Society for Nutritional Sciences Journal of Nutrition* 131 (2001): 409S–420S.

50. Kline, *Bodies of Knowledge*, 2.

51. League supporter and medical adviser Herbert Ratner was a longtime advocate of the idea that "anatomy is destiny," a point he articulated through-

out his involvement with the organization. Ratner to Froehlich (23 July 1971), Folder—Herbert Ratner, Box 81, Physician Files, LLLIR.

52. Adrienne Rich, *Of Woman Born* (New York: W. W. Norton, 1976), 39.

53. Ibid., 39; Kline, *Bodies of Knowledge*, 139. Kline reports that the number of out-of-hospital births "more than doubled, from 0.6 percent in 1970 to 1.5% in 1977, mostly attributable to the increasing popularity of planned home births."

54. Boston Women's Health Book Collective (BWHBC), *Our Bodies, Ourselves* (New York: Simon and Schuster, 1973), 2.

55. Kline, *Bodies of Knowledge*, 129. Kline writes that in the 1970s, mothers who sought empowering childbirth experiences not only encountered resistance within the medical system but "encountered resistance from a divisive feminist community reluctant to embrace a philosophy that had its roots in a conservative celebration of motherhood." See also Wolf, "What Feminists Can Do," 397–424.

56. Agnew, *Back from the Land*, 179.

57. BWHBC, *Our Bodies*, 311.

58. Ibid.

59. Ibid.

60. Ibid.

61. Several of the women whom I interviewed, for example, said they relied heavily on the book *The Complete Book of Breastfeeding* (New York: Workman, 1972), by Marion S. Eiger and Sally Wendkos Olds.

62. BWHBC, *Our Bodies*, 312.

63. *Our Bodies, Ourselves* did originally contain some basic information on breastfeeding.

64. Sandra Morgen, *Into Our Own Hands: The Women's Health Movement in the United States, 1969–1990* (New Brunswick, NJ: Rutgers University Press, 2002).

65. Mary K. White, "What's in It for Mother?" *La Leche League News* 8, no. 5 (September–October 1966): 7–10.

66. BWHBC, *Our Bodies*, 12–13.

67. Weiner, "Reconstructing Motherhood," 1381.

68. Letter from M. A., 11 July 1972, Folder 14, Box 52, BWHBC Records, SLSC.

69. Rich, *Of Woman Born*, 22.

70. Ibid.

71. Transcript from *The Phil Donahue Show*—07311, Folder 15, Box 41, BWHBC Records, SLSC.

72. Ibid.

73. Ibid.

74. *A La Leche League Dialogue: An Historic Document* [reprint], *Child and Family*, 1981, p. 32, Box 10, Publications Files, LLLIR. In 1956, six of the origi-

nal La Leche League founders participated in a dialogue moderated by Herbert Ratner with the purpose of clarifying their organization's mission. In the context of this dialogue, one of the founders stated, "We don't think [the mother] should be working," a statement echoed by at least four of those present. Another founder stated, "We'd show that it was necessary for the best development of her children and herself and the family life that she stay with them."

75. Transcript from *The Phil Donahue Show.*

76. Ibid.

77. Ibid.

78. Letter to BWHBC, 7 August 1981, Folder 15, Box 41, BWHBC, Schlesinger Library.

79. Interview with Patty (2 February 2009), author's files.

80. Letter to BWHBC, 12 August 1981, Folder 15, Box 41, BWBHC, Schlesinger Library.

81. Interview with Tammy (11 February 2009), author's files.

82. Letter to LLLI Board of Directors, 13 July 1978, Folder 1, Box 7, Board of Directors Files, LLLIR.

83. Letter from Edwina Froehlich, 12 September 1978, Folder 1, Box 7, Board of Directors Files, LLLIR.

84. Board member "Nancy" to Edwina Froehlich, 22 July 1978, Folder 1, Box 7, LLLIR.

85. Ibid.

86. "Outreach: Lulu Leader Faces the BIG WIDE WORLD," LLLI internal handbook (n.d.), Box 10, Publications Files, LLLIR.

87. "Background for Leadership Preparation," *Leaven* 5, no. 5 (September–October 1969), front page.

88. *La Leche League Dialogue*, 11.

89. Letter to Board, August 1980, Folder 1, Box 7, Board of Directors Files, LLLIR.

90. Ibid.

91. Letter from Edwina, 30 July 1977, Folder 1, Box 7, Board of Directors Files, LLLIR.

92. Ibid.

93. "Background for Leadership Preparation," front page.

94. This is a reference to the "Marian Ideal" prevalent in Catholicism throughout its history and the model of imperfection it has required. It speaks to what many have observed is an unrealistic and unattainable model of womanhood which Mary, the virgin mother, has provided for female followers. See, for example, Carolyn Osiek, *Beyond Anger: On Being a Feminist in the Church* (New York: Paulist Press, 1986). She observed that Mary offers an "impossible ideal to which no woman could attain, with whom all women are invited to feel inadequate" (20).

95. Letter to Board, 19 July 1977, Folder 1, Box 7, Board of Directors Files, LLLIR.

96. Leader in Indiana to Board, 12 October 1978, Folder 1, Box 7, Board of Directors Files, LLLIR.

97. Letter from Edwina Froehlich, 2 November 1978, Folder 1, Box 7, Board of Directors Files, LLLIR.

98. La Leche League, "Announcing New Third Concept," *Leaven* 15, no. 6 (November–December 1979).

99. Husband of an LLL leader to Marian Thompson, 3 November 1979, Folder 1, Box 7, Board of Directors Files, LLLIR.

100. "La Leche League International Policies and Standing Rules" (1984), Box 10, Publications Files, LLLIR.

101. Board Response Letter, 7 April 1980, Folder 1, Box 7, Board of Directors Files, LLLIR.

102. Letter to Board, 5 July 1980, Folder 1, Box 7, Board of Directors Files, LLLIR.

103. Interview with Dolly (17 February 2009), author's files.

104. Ibid.

105. Ibid.

Chapter 6

1. "The Baby Won't Suck?" *La Leche League News* 4, no. 2 (November 1961): 1, 6.

2. Ibid.

3. Earl M. Tarr, "Development and Re-establishment of Breast Milk by Use of Dr. Abt's Electric Breast Pump," *California and Western Medicine* 23, no. 6 (June 1925): 728–32.

4. Calvina MacDonald, "Abt's Electric Breast Pump," *American Journal of Nursing* 25, no. 4 (April 1925): 277–80.

5. Swanson, "Human Milk as Technology," 20–36.

6. See, for example, Kate Boyer and Maia Boswell-Penc, "Breast Pumps: A Feminist Technology or (Yet) 'More Work for Mother'?" in *Feminist Technology*, ed. Linda L. Layne, Sharra L. Vostral, and Kate Boyer (Urbana-Champaign: University of Illinois Press, 2010), 119–35.

7. See Leavitt, *Brought to Bed*; Martin S. Pernick, *A Calculus of Suffering: Pain, Professionalism and Anesthesia in Nineteenth-Century America* (New York: Columbia University Press, 1985); Sandelowski, *Pain, Pleasure, and American Childbirth*; and Wolf, *Deliver Me from Pain*.

8. Audrey Davis and Toby Appel, *Bloodletting Instruments in the National*

Museum of History and Technology, Smithsonian Studies in History and Technology no. 41 (Washington, DC: Smithsonian Institute Press, 1979), 32.

9. Ibid., 26.

10. Ibid.

11. Ibid., 34.

12. Ibid.

13. Ibid., 86.

14. Tilio Suzzara Verdi, *Maternity: A Popular Treatise for Young Wives and Mothers* (New York: J. B. Ford, 1873), excerpted and reprinted in "Time Was," *Journal of Human Lactation* 5, no. 3 (1989): 138–42, quotations on 138.

15. Davis and Appel, *Bloodletting*, 34.

16. Patent No. 1,644,257, "Breast Pump," United States Patent Office, issued to Edward Lasker of Chicago, IL, 4 October 1927. (Application filed 2 August 1923—granted 1927.)

17. Abt's 85th Birthday Speech (18 December 1952), Folder 1, Box 1, Series no. 18/3/14/2, Isaac Arthur Abt Papers, Northwestern University Archives, Evanston, IL; Arthur F. Abt, "Obituary—Isaac Arthur Abt," *Pediatrics* 18, no. 2 (August 1956): 327–35.

18. Isaac A. Abt, *The Baby Doctor* (New York: McGraw-Hill, 1944), 251.

19. Ibid., 252.

20. MacDonald, "Abt's Electric Breast Pump," 279.

21. Abt, *Baby Doctor*, 252.

22. Tarr, "Development and Re-establishment," 728.

23. Edward Lasker, *Chess Secrets I Learned from the Masters* (London: Hollis and Carter, 1952), 249–50. Lasker is best known for his status as an international chess champion. After his friends nicknamed him the "chest player," he perhaps decided to downplay his role in the development of breast pump technology, despite his own acknowledgment that this invention provided him with a lifetime of financial security.

24. Joel D. Howell, *Technology in the Hospital: Transforming Patient Care in the Early Twentieth Century* (Baltimore: Johns Hopkins University Press, 1995), 18.

25. E-mail from Bjorn Bernhard Englund (21 May 2010), personal collection of Rolf Egnell, Einar Egnell's grandson, author's files.

26. Letter from Olof Levan (10 June 2010), personal collection of Rolf Egnell, author's files.

27. The information on Egnell's development of the breast pump is based on the following sources: "Einar Egnell" [film] provided courtesy of Ameda, Inc., author's files, (excerpt available online at http://vimeo.com/17615133 [accessed 7 December 2014]); letter from Egnell's employee, Olof Levan (19 November 2012), author's files; Hjalmar Cederström, *Södersjukhuset, Stockholm* (Stockholm, Sweden: Rydahl, 1946), 1–67; Ollë Larsson, *The Life of the Swedish-Swiss*

Entrepreneur: An Autobiography (Zug, Switzerland: Ollë Larsson Holding, 2009), copy provided courtesy of Medela, Inc.; Jan Riordan and Karen Wambach, eds., *Breastfeeding and Human Lactation*, 4th ed. (Sudbury, MA: Jones and Bartlett Publishers, 2010), 387–88.

28. Einar Egnell, "The Mechanics of Different Methods of Emptying the Female Breast" [translated and shortened], *Svenska Lakartidningen* [Journal of the Swedish Medical Association] 40 (October 1956), copy provided courtesy of Ameda, Inc.

29. Niles Newton to Edwina Froehlich, 27 September 1965, Folder—Niles Newton, Box 80, Physician Files, LLLIR.

30. Ibid. Enclosed pamphlet: "Egnell's Breast Pump" (c.1959).

31. Betty Wagner to Niles Newton, 8 October 1965, Folder—Niles Newton, Box 80, Physician Files, LLLIR.

32. Bobel, "Bounded Liberation," 130–51, 135.

33. "Fashions for Subtle Nursing," *La Leche League News* 2, no. 2 (July–August 1959), 4; "These clothes for the nursing mother . . ." *La Leche League News* 5, no. 4 (July–August 1963): 2; "Sewing Kits for the Nursing Mother," *La Leche League News* 7, no. 1 (January–February 1965): 4; "Happy Baby Carriers, Grinderoo, Comfy-Dry Pads, Happy Nursing Bras, One Hand Fastnurse, Natural Look" [fundraising advertisement], *Why Breastfeed Your Baby?* (Franklin Park, IL: La Leche League International, 1980), back page.

34. La Leche League literature often focused on how to nurse discreetly. See, for example, Marion Thompson, "Vacationing with Baby," *La Leche League News* 2, no. 1 (May–June 1959): 4.

35. Letter to Spock, 29 March 1960, Folder March 1960, Box 6, Group 1, BSPP. In her letter, one woman recounted at length her pregnancy, childbirth, and postpartum experiences including details on her use of a nipple shield to help her.

36. Although I do not believe there is one answer to the question, Is the breast pump a feminist technology? My perspective on what constitutes a feminist technology is one that positions women in an active role in the production and/or use of the device, rather than a passive one. For more on the question of what constitutes a feminist technology, see Linda L. Layne, "Introduction," in *Feminist Technology*, ed. Linda L. Layne, Sharra L. Vostral, and Kate Boyer (Urbana: University of Illinois Press, 2010), 1–35; and Ruth Oldenziel, "Man the Maker, Woman the Consumer: The Consumption Junction Revisited," in *Feminism in Twentieth-Century Science, Technology, and Medicine* (Chicago: University of Chicago Press, 2001), 128–48.

37. Donna Haraway, "A Cyborg Manifesto: Science, Technology, and Socialist-Feminism in the Late Twentieth Century," in *The International Handbook of Virtual Learning Environments*, ed. J. Weiss (Dordrecht, The Netherlands: Springer, 2006), 117–58, 117.

38. Ronald Kotulak, "Mom's Return to Job Called Peril to Baby: Expert Tells Need of Mothering in 1st Year," *Chicago Tribune* (14 November 1969) [news clipping], Folder 3, Box 1, Betty Wagner Files, LLLIR; Barbara Jacobs, "The Revival of Breast-Feeding," *East West Journal* 8, no. 2 (February 1978) [reprint no. 144 June 1979, La Leche League], Folder 3, Box 1, Marian Thompson, LLLIR.

39. On tinkering, see Susan Douglas, *Inventing American Broadcasting, 1899–1922* (Baltimore: Johns Hopkins University Press, 1987); Kathleen Franz, *Tinkering: Consumers Reinvent the Early Automobile* (Philadelphia: University of Pennsylvania Press, 2005); and Rachel Plotnick, "At the Interface: The Case of the Electric Push Button, 1880–1923," *Technology and Culture* 53, no. 4 (October 2012): 815–45.

40. Judy Torgus, "Breast Pumps," *Leaven* 6, no. 3 (May–June 1970): 8.

41. Ibid.

42. Mary Ann Cahill, "Breastfeeding and Working," Publication No. 58 (Franklin Park, IL: La Leche League International, 1976), Box 10, Publications Files, LLLIR.

43. Faith Bedford, "From ILCA's President," *Journal of Human Lactation* 1, no. 2 (1985): 16–17.

44. Kathleen Auerbach, "Timing of IBLCE and ILCA," LACTNET [Listserv] (29 December 1996), http://community.lsoft.com/scripts/wa-LSOFTDONATIONS.exe?A0=LACTNET (accessed 24 September 2013).

45. Kathleen Auerbach, "A We/Them Dichotomy," *Journal of Human Lactation* 5, no. 3 (1989): 121.

46. Ibid.

47. Linda J. Smith, "Origins of ILCA and IBLCE," LACTNET [Listserv] (30 December 1996), http://community.lsoft.com/scripts/wa-LSOFTDONATIONS.exe?A0=LACTNET (accessed 24 September 2013).

48. Ibid.

49. Jan Barger, "President's Letter: Each One Reach One," *Journal of Human Lactation* 6, no. 3 (1990): 148–49.

50. Bedford, "From ILCA's President," 16.

51. Barbara Bono, "Letter Received and Invited Reply," *Journal of Human Lactation* 5, no. 1 (1989): 37.

52. "Biographical Sketches of Candidates for Office," *Journal of Human Lactation* 1, no. 1 (1985): 11–13.

53. Bedford, "From ILCA's President," 16–17.

54. Ibid.

55. "IBLCE News," *Journal of Human Lactation* 1, no. 2 (1985): 24.

56. Jeanine Klaus, "Some Thoughts on the ILCA Conference," *Journal of Human Lactation* 2, no. 2 (1986): 79.

57. Chris Mulford, "Women and Power," *Journal of Human Lactation* 2, no. 2 (1986): 54–55, quotation on 55.

58. On woman-run women's health care, see Kline, *Bodies of Knowledge*, 65–96. On the interaction between women's health care workers and mainstream medicine, see Morgen, *Into Our Own Hands*, 131–52.

59. Kathleen Auerbach, "Blaming the Victim," *Journal of Human Lactation* 3, no. 3 (1987): 80–81.

60. Ibid.

61. Ibid.

62. "News of Local Affiliates—Eastern Pennsylvania," *Journal of Human Lactation* 2, no. 2 (1986): 69–70; Mary Bachman, "Lactation Consultant in a Hospital Setting," *Journal of Human Lactation* 3, no. 3 (1987): 104–5.

63. Jan Berger and Judith Lawuers, "Reply from ILCA President and 1990 Conference Convenor," *Journal of Human Lactation* 6, no. 4 (1990): 160–61.

64. Kathleen Auerbach, "Editor's Note—Besides Yourself, What Else Are You Selling?" *Journal of Human Lactation* 5, no. 4 (1989): 168–69, quotation on 169.

65. See, for example, Jeanne Lambert, "Dear Editor," *Journal of Human Lactation* 6, no. 4 (1990): 159; Alison Hazelbaker, "Technology and Breastfeeding," *Journal of Human Lactation* 7, no. 3 (1991): 174–75, 174; Linda Shrago, "Fostering Collegial Relationships among Lactation Consultants," *Journal of Human Lactation* 11, no. 1 (1995):1–2, 1.

66. Jack Newman, "Breastfeeding Problems Associated with the Early Introduction of Bottles and Pacifiers," *Journal of Human Lactation* 6, no. 2 (1990): 59–63.

67. Hazelbaker, "Technology and Breastfeeding," 174–75; J. F., "Re: Nipple Confusion," LACTNET [Listserv] (20 May 1995), http://community.lsoft.com/scripts/wa-LSOFTDONATIONS.exe?A0=LACTNET (accessed 2 October 2013). Wherever I discuss a posting on LACTNET, I am referring to an archived post visible to the general public online. Additional conversations, responses, and so forth are not visible to nonmembers and have not been used in my analysis. Out of respect for the membership's privacy, I have not utilized any information that is not accessible by the general public.

68. Kathleen Auerbach, "Breastmilk versus Breastfeeding: Product versus Process," *Journal of Human Lactation* 7, no. 3 (1991): 115–16.

69. Although opinions varied widely, LCs tended to push for exclusive breastfeeding for as long as possible, with the consensus on when introducing a bottle would be least likely to harm an infant's latch hovering somewhere between four to eight weeks. Search term "introduce bottle" on LACTNET [Listserv], http://community.lsoft.com/scripts/wa-LSOFTDONATIONS.exe?A0=LACTNET) (accessed 6 December 2013).

70. J. D., "Re: Nipple Confusion." LACTNET [Listserv] (18 July 1995), http://community.lsoft.com/scripts/wa-LSOFTDONATIONS.exe?A0=LACTNET (accessed 26 September 2013).

71. Newman, "Breastfeeding Problems," 59–63; Y. S., "Cup Feeding," LACTNET [Listserv] (20 June 1995), http://community.lsoft.com/scripts/wa-LSOFTDONATIONS.exe?A0=LACTNET (accessed 2 October 2013).

72. L. A., "Bottle Teats—Hey, Let's Talk about Them," LACTNET [Listserv] (18 July 1995), http://community.lsoft.com/scripts/wa-LSOFTDONATIONS.exe?A0=LACTNET (accessed 26 September 2013).

73. Gabrielle Palmer, "Give Breastfeeding Back to Mothers," *Journal of Human Lactation* 7, no. 1 (1991): 1–2, quotation on 1.

74. Ibid., 2.

75. Auerbach, "Breastmilk versus Breastfeeding," 115.

76. Ibid., 116.

77. Palmer, "Give Breastfeeding Back," 1.

78. Susan Switzer, "Letters—Breast Milk Feeding Is Not Breastfeeding," *Journal of Human Lactation* 7, no. 3 (1991): 117–18, quotation on 117.

79. Ibid., 117–18.

80. Judy Hopkinson, "Letters—LCs, Remember Your Roots!" *Journal of Human Lactation* 7, no. 3 (1991): 117.

81. Hazelbaker, "Technology and Breastfeeding," 175.

82. Kate Yates Pierce and Mary Rose Tully, "Mother's Own Milk: Guidelines for Storage and Handling," *Journal of Human Lactation* 8, no. 3 (1992): 159–60.

83. Victor Oliveira, Elizabeth Racine, Jennifer Olmsted, and Linda M. Ghelfi, "The WIC Program: Background, Trends, and Issues," *Food Assistance and Nutrition Research Report No. 27*, Food and Rural Economics Division, Economic Research Service, U.S. Department of Agriculture (October 2002), available online at http://www.ers.usda.gov/publications/fanrr-food-assistance-nutrition-research-program/fanrr27.aspx (accessed 30 September 2013), 7.

84. Ibid., 2, 4. For a discussion of the impact of WIC on breastfeeding more generally, see Diane Thulier, "Breastfeeding in America: A History of Influencing Factors," *Journal of Human Lactation* 25, no. 1 (2009): 85–94.

85. Oliveira et al., "WIC Program," 9–10. There is a special irony in the reality that the federal government will fund breastfeeding technologies for economically disadvantaged mothers while these same women confront some of the worst employment conditions in positions that do not allow for paid maternal leave, paid breaks for breast pumping, or workplace support for the practice more generally. As of 23 March 2010, Section 7(1)-(4) of the Fair Labor Standards Act of 1938 (29 U.S.C. 207) contained protections for nursing mother employees, including providing time and a designated space (not a bathroom) for breast pumping. The amended clauses, however, also exempted business with fewer than fifty employees and did not require employers to pay employees for the breaks.

86. "Report from the United States APHA Committee on WIC," *Journal of Human Lactation* 5, no. 4 (1989): 200.

87. Nancy Schweers, "Dear Editor," *Journal of Human Lactation* 6, no. 1 (1990): 3–4, quotation on 4.

88. Ruth Forni, "Dear Editor," *Journal of Human Lactation* 6, no. 3 (1990): 103.

89. Susan Switzer, "Letters—Breast Milk Feeding Is Not Breastfeeding," 117–18; "60% of U.S. Families Have 2 Wage Earners," *Chicago Tribune* (1 November 1979) [clipping], Folder 5, Box 1, Betty Wagner Files, LLLIR. A 1981 report sponsored by the Rand Corporation showed that between 1950 and 1980, the percentage of married mothers with children under six in the labor force rose from 12 percent to 45 percent, Linda J. Waite, "U.S. Women at Work" (Santa Monica, CA: Rand Corporation, 1981), 3.

90. Here I am situating breastfeeding technology within a broader historical narrative on the commodification of countercultural politics and ideologies in the late twentieth century. See Andrew Kirk, "Appropriating Technology: The Whole Earth Catalog and Counterculture Environmental Politics," *Environmental History* 6, no. 3 (July 2001): 374–94.

91. Susan Aldana, "Dear Editor," *Journal of Human Lactation* 6, no. 2 (1990): 51.

92. Auerbach, "Besides Yourself," 168.

93. Cherie Stock, "Breastfeeding Promotion in a Rural Indiana WIC Office," *Journal of Human Lactation* 8, no. 3 (1992): 172–79; Sharon Breunig et al., "The Florida Breastfeeding Promotion Project: A Coalition Effort to Improve Hospital Practices and Policies," *Journal of Human Lactation* 8, no. 4 (1992): 213–15; Jane Bagwell et al., "Breastfeeding among Women in the Alabama WIC Program," *Journal of Human Lactation* 8, no. 4 (1992): 205–8; Elaine Nadel, "Breastfeeding Promotion in an Urban New Jersey WIC Office," *Journal of Human Lactation* 9, no. 2 (1993): 140–42; Marianne Michaels, "Breastfeeding Promotion in the Utah WIC Program," *Journal of Human Lactation* 9, no. 3 (1993): 206–7; Deborah Myers et al., "The Joys and Challenges of Promoting Breastfeeding in South Central Los Angeles," *Journal of Human Lactation* 9, no. 4 (1993): 284–87.

94. Stock, "Breastfeeding Promotion," 178.

95. Ibid.

96. "Lactnet Empowers the Frontlines of Breastfeeding Education, One Email at a Time: Global Lactation Support Community Wins Top Honour in L-Soft's Listserv Choice Awards," L-Soft press release (20 October 2006), available online at http://www.lsoft.com/news/2006/choice2005-5-uk.asp (accessed 20 August 2014).

97. Ibid.

98. J. F., "Pump Characteristics," LACTNET [Listserv] (24 February 1997), http://community.lsoft.com/scripts/wa-LSOFTDONATIONS.exe?A0=LACTNET (accessed 3 October 2013).

99. P. B., "Too Many Pump Rentals," LACTNET [Listserv] (2 October 1995), http://community.lsoft.com/scripts/wa-LSOFTDONATIONS.exe?A0=LACTNET (accessed 3 October 2013).

100. D. S., "Hospital Giving Breastpumps," LACTNET [Listserv] (6 May 1996), http://community.lsoft.com/scripts/wa-LSOFTDONATIONS.exe?A0=LACTNET (accessed 2 October 2013).

101. M. W., "FU on Pumpy Love," LACTNET [Listserv] (11 January 1996), http://community.lsoft.com/scripts/wa-LSOFTDONATIONS.exe?A0=LACTNET (accessed 6 December 2013).

102. The task force included Alvin M. Mauer, Lewis A. Barness, L. J. Filer, Frederick C. Holmes, and William B. Weil. "The Promotion of Breast-Feeding: Policy Statement Based on Task Force Report," *Pediatrics* 69, no. 5 (1982): 654–61.

103. "International Code of Marketing of Breast-Milk Substitutes" (Geneva: World Health Organization, 1981). In 1974 the British group War on Want published a pamphlet called "The Baby Killer," exposing the devastation caused by the marketing, sale, and use of infant formula in the developing world. The pamphlet raised international attention to the issue and in 1977 the group Infant Formula Action Coalition (INFACT) began a consumer boycott against Nestlé, the largest and most visible international peddler of commercial baby foods. In 1978, public hearings took place under the oversight of Senator Ted Kennedy on the issue and in 1979, the WHO and UNICEF met on the subject of infant and young child feeding. As a result of these events, the WHO published its Code of Marketing Breast-Milk Substitutes and later led multiple international efforts to support breastfeeding for infants around the globe. For a detailed discussion on the history of the Nestlé boycott, see John Dobbing, ed., *Infant Feeding: Anatomy of a Controversy 1973–1984* (New York: Springer-Verlag, 1988).

104. Mauer et al., "Promotion of Breast-Feeding," 654.

105. *Report of the Surgeon General's Workshop on Breastfeeding and Human Lactation*, U.S. Department of Health and Human Services (11–12 June 1984), DHHS Publication no. HRS-D-MC 84–2, p. iii.

106. Ibid.

107. In 1985, Ruth Lawrence founded and directed the Breastfeeding and Human Lactation Study Center at the University of Rochester. She went on to serve as a member of the American Academy of Pediatrics' Work Group on Breastfeeding as well as the U.S. National Breastfeeding Committee, and she is a founding member of the Academy of Breastfeeding Medicine. See Ruth A. Lawrence, *Breastfeeding: A Guide for the Medical Profession*, 7th ed. (Maryland Heights, MO: Mosby/Elsevier), 2011.

108. *Report of the Surgeon General's Workshop*, iv–v.

109. Ibid., 67.

110. Ibid., 69.

111. Ibid., 70.

112. "Innocenti Declaration: On the Protection, Promotion and Support of Breastfeeding," (30 July 30–1 August 1990), available online at http://www.unicef.org/programme/breastfeeding/innocenti.htm (accessed 20 August 2014).

113. "Summary of Minutes, AAP Work Group on Breastfeeding," 15–16 November 1997, Folder—Work Group on Breastfeeding Board Action, Box 361, Pediatric History Center Collection, American Academy of Pediatrics, Elk Grove Village, IL (hereafter PHCC).

114. "Minutes of the Advisory Committee to the Board on the Office of Community Pediatrics," 22 May 1994, Folder—Work Group on Breastfeeding Board Action, Box 361, PHCC.

115. AAP Board of Directors Meeting, 20 October 1994, Folder—Work Group on Breastfeeding Board Action, Box 361, PHCC.

116. "Summary of Minutes Work Group on Breastfeeding," 8–9 February 1997, Folder—Work Group on Breastfeeding Board Action, Box 361, PHCC.

117. Abigail Trafford, "What's Good for the Baby May Guilt-Trip the Mother," *Washington Post* (9 December 1997) [clipping], Folder—Publications, Serials, Breastfeeding: Best for Babies, January, April, Box 563, PHCC.

118. Margo Slade, "Have Pump, Will Travel," *New York Times* (14 December 1997) [clipping], Folder—Publications, Serials, Breastfeeding: Best for Babies, January, April, Box 563, PHCC.

119. Ibid.

120. Ibid.

121. L. John Horwood and David M. Fergusson, "Breastfeeding and Later Cognitive and Academic Outcomes," *Pediatrics* 101, no. 1 (1 January 1998): e9.

122. *Talk of the Nation*, National Public Radio (19 January 1998), Folder—Committees, Work Group on Breastfeeding, Radio Program Transcript 1998, Box 361, PHCC.

123. Ibid.

124. Ibid.

125. Ibid.

126. Barbara Behrmann, *The Breastfeeding Café: Mothers Share the Joys, Challenges, and Secrets of Nursing* (Ann Arbor: University of Michigan Press, 2005), 167.

127. Ibid., 168.

Epilogue

1. *Newsweek* coined the term "mommy wars" in 1990, and by end of the century it was common parlance. See "Mommy vs. Mommy," *Newsweek* (3 June 1990), available online at http://www.newsweek.com/mommy-vs-mommy

-206132 (accessed 7 December 2014); Tracy Thompson, "A War inside Your Head," *Washington Post* (15 February 1998), W12, available online at http://www.washingtonpost.com/wp-srv/national/longterm/mommywars/mommy.htm (accessed 7 December 2014); and Susan Douglas and Meredith Michaels, "The Mommy Wars: How the Media Turned Motherhood into a Catfight," *Ms. Magazine* (February–March 2000), available online at http://www.msmagazine.com/feb00/mommywars1.asp (accessed 7 December 2014). For more on the etiology of "mommy wars," see Katherine L. Eaves, "Where Is the Love? How Parenting Magazines Discuss the 'Mommy Wars' and Why It Matters," *Proceedings of the 4th Annual GRASP Symposium* (Wichita State University, 2008), available online at http://soar.wichita.edu:8080/bitstream/handle/10057/1344/grasp-2008-12.pdf?sequence=1 (accessed 7 December 2014).

2. Hanna Rosin, "The Case against Breast-Feeding," *Atlantic* (1 April 2009), available online at http://www.theatlantic.com/magazine/archive/2009/04/the-case-against-breast-feeding/307311/ (accessed 20 August 2014).

3. Ibid.

4. Lisa Belkin, "Is Breastfeeding the New Vacuum Cleaner?" *New York Times* (16 March 2009), available online at http://parenting.blogs.nytimes.com/2009/03/16/is-breastfeeding-the-new-vacuum-cleaner/?_r=0 (accessed 20 August 2014). See especially the "comments" section at the bottom of the page.

5. Danielle Bridges, "Breastfeeding Haiku" (27 July 2011), available online at http://kellymom.com/blog-post/breastfeeding-haiku/ (accessed 20 August 2012).

6. The first objectives targeted the year 1990. Subsequent versions have focused on a specific set of public health goals to be met at ten-year intervals. Increasing breastfeeding has been a part of overall efforts to improve maternal and child health. See http://www.healthypeople.gov/2020/about/history.aspx (accessed 20 August 2014); and Philip J. Hilts, "Nation Is Falling Short of Health Goals for 2000," *New York Times* (11 June 1999), A30.

7. According to a 2012 American Academy of Pediatrics (AAP) statement, the CDC draws its data on breastfeeding from the National Immunization Survey, Findings from the National Health and Nutrition Examination Surveys, and National Survey of Maternity Practices in Infant Nutrition and Care. See AAP Section on Breastfeeding, "Breastfeeding and the Use of Human Milk," *Pediatrics* 129 (27 February 2012), e827–e841, e828.

8. Pamela Hill, "Update on Breastfeeding: Healthy People 2010 Objectives," *American Journal of Maternal/Child Nursing* 25, no. 5 (September–October 2000): 248–51; Department of Health and Human Services, Centers for Disease Control and Prevention, *Breastfeeding Report Card—United States, 2010* (August 2010), available online at http://www.cdc.gov/breastfeeding/pdf/BreastfeedingReportCard2010.pdf (accessed 20 August 2014). Data for the CDC breastfeeding report card comes from the National Immunization Survey (NIS), conducted by the National Center for Immunizations and Respiratory

Diseases (NCIRD) and the National Center for Health Statistics, Centers for Disease Control and Prevention. It is a "list-assisted random digit-dialing telephone survey followed by a mailed survey to children's immunization providers that began data collection in April 1994." More information is available online at http://www.cdc.gov/nchs/nis.htm (accessed 20 August 2014).

9. Section on Breastfeeding, "Breastfeeding," e828.

10. CDC National Immunization Survey, *Breastfeeding among U.S. Children Born 2000–2010* (last updated 31 July 2013), available online at http://www.cdc.gov/breastfeeding/data/NIS_data/index.htm (accessed 20 August 2014).

11. CDC, "Progress in Increasing Breastfeeding and Reducing Racial/Ethnic Differences—United States, 2000–2008 Births," *MMWR* 62, no. 5 (8 February 2013): 77–80. The often neglected history of black women working as wet nurses for white children throughout slavery and well into the twentieth century has undoubtedly lent a host of conflicting meanings to breastfeeding in the African American community. During Women's History Month in 2014 the subject of the history of black wet nurses in relation to contemporary rates of breastfeeding among black women prompted a lively discussion on Twitter. See "The History of Breastfeeding among Black Women—What White Nurses Need to Know," *ofcourseitsaboutyou* [blog] (10 March 2014), available online at http://ofcourseitsaboutyou.com/2014/03/10/the-history-of-breastfeeding-among-black-women-what-white-nurses-need-to-know/ (accessed 5 August 2014).

12. Erika C. Odom, Ruowei Li, Kelly S. Scanlon, Cria G. Perrine, and Laurence M. Grummer-Strawn, "Reasons for Earlier Than Desired Cessation of Breastfeeding," *Pediatrics* 131, no. 3 (March 2013): e726–e732; Indu B. Ahluwalia, Brian Morrow, and Jason Hsia, "Why Do Women Stop Breastfeeding? Findings from the Pregnancy Assessment and Monitoring System," *Pediatrics* 116, no. 6 (December 2005): 1408–12.

13. Li Ruowei, Sara B. Fein, Jian Chen, and Laurence M. Grummer-Strawn, "Why Mothers Stop Breastfeeding: Mothers' Self-Reported Reasons for Stopping during the First Year," *Pediatrics* 122, supplement 2 (October 2008): S69–S76, S69.

14. Ibid.

15. CDC National Immunization Survey, "NIS Survey Methods" (last updated 31 July 2013), available online at http://www.cdc.gov/breastfeeding/data/NIS_data/survey_methods.htm (accessed 20 August 2014).

16. Sheela R. Geraghty and Kathleen M. Rasmussen, "Redefining 'Breastfeeding' Initiation and Duration in the Age of Breastmilk Pumping," *Breastfeeding Medicine* 5, no. 3 (2010): 135–37, quotation on 135.

17. Ibid.

18. Section on Breastfeeding, "Breastfeeding," e827–e841.

19. Neal Conan and Jill Lepore, *Talk of the Nation: The Politics of the Breast Pump*, radio broadcast on National Public Radio (14 January 2009), available

online at http://www.npr.org/templates/story/story.php?storyId=99353070 (accessed 20 August 2014).

20. See, for example, "Bad Lactation Consultant Experience—What Did You Do?" (24 May 2010), online forum on *DC Urban Mom*, available online at http://www.dcurbanmom.com/jforum/posts/list/15/107811.page (accessed 20 August 2014); "Bad Lactation Consultant" (24 April 2013), online forum on *The Bump*, available online at http://forums.thebump.com/discussion/8959872/bad-lactation-consultant (accessed 20 August 2014).

21. Asian Communities of Reproductive Justice, *A New Vision for Advancing Our Movement for Reproductive Health, Reproductive Rights and Reproductive Justice* (2005), 3, available online at http://strongfamiliesmovement.org/assets/docs/ACRJ-A-New-Vision.pdf (accessed 5 August 2014).

22. Miriam Zoila Pérez, "An Open Letter to the New Times and the Reproductive Rights Movement," *Colorlines.com* (31 July 2014), available online at http://colorlines.com/archives/2014/07/an_open_letter_to_the_new_york_times_race_and_the_reproductive_rights_movement.html (accessed 5 August 2014).

23. Asian Communities of Reproductive Justice, *New Vision*, 4.

24. Jill Lepore, "If Breast Is Best, Why Are Women Bottling Their Milk?" *New Yorker* (19 January 2009), available online at http://www.newyorker.com/reporting/2009/01/19/090119fa_fact_lepore?printable=true (accessed 14 December 2013).

25. For more on the nature of this debate over the validity of scientific evidence of breastfeeding's superiority, see Joan B. Wolf, "Is Breast Really Best? Risk and Total Motherhood in the National Breastfeeding Awareness Campaign," *Journal of Health Politics, Policy and Law* 32, no. 4 (August 2007): 595–636; Judy M. Hopkinson, "Response to 'Is Breast Really Best? Risk and Total Motherhood in the National Breast-Feeding Awareness Campaign,'" *Journal of Health Politics, Policy and Law* 32, no. 4 (August 2007): 637–48; Joan B. Wolf, "Rejoinder to Judy M. Hopkinson," *Journal of Health Politics, Policy and Law* 32, no. 4 (August 2007): 649–54; and Joan B. Wolf, *Is Breast Best? Taking on the Breastfeeding Experts and the New High Stakes of Motherhood* (New York: New York University Press, 2011).

26. See Penny Van Esterik, "Breastfeeding and Feminism," *International Journal of Gynecology and Obstetrics* 47 (1994): S41–S54, S48; and Hausman, *Mother's Milk*, 216–28.

27. See, for example, Suzanne Barston, *Bottled Up: How the Way We Feed Babies Has Come to Define Motherhood, and Why It Shouldn't* (Chicago: University of Chicago Press, 2012), and her *Fearless Formula Feeder* [blog], available online at http://www.fearlessformulafeeder.com (accessed 20 August 2014). While the argument of critic, author, journalist, and mother Suzanne Barston that women should be free to feed their children based on accurate information

and without judgment is admirable, feminist, and valid, she also places an emphasis on infant feeding "choice" that I argue is still not actually available to the majority of mothers in the United States at this time.

28. For examples of breastfeeding expectations and guilt, see Jane E. Brody, "The Ideal and the Real of Breast-Feeding," *New York Times* (23 July 2012), available online at http://well.blogs.nytimes.com/2012/07/23/the-ideal-and-the-real-of-breast-feeding/ (accessed 20 August 2014); Alissa Quart, "The Milk Wars," *New York Times Sunday Review* (14 July 2012), available online at http://www.nytimes.com/2012/07/15/opinion/sunday/the-breast-feeding-wars.html (accessed 20 August 2014); Lisa Belkin, "In Support of Bottle-Feeding," *New York Times* (22 July 2009), available online at http://parenting.blogs.nytimes.com/2009/07/22/in-support-of-bottle-feeding/?gwh=56B9FE120F97A61213143868D4F473C9 (accessed 20 August 2014); and Tara Parker-Pope, "Most Moms Give Up on Breastfeeding," *New York Times* (11 August 2008), available online at http://well.blogs.nytimes.com/2008/08/11/most-moms-give-up-on-breast-feeding/ (accessed 20 August 2014) (see especially the "Comments" section).

29. See Emily Bazelon, "Milk Me: Is the Breast Pump the New BlackBerry?" *Slate* (27 March 2006), available online at http://www.slate.com/articles/life/family/2006/03/milk_me.html (accessed 20 August 2014); and Maia Boswell-Penc and Katherine Boyer, "Expressing Anxiety? Breast Pump Usage in American Wage Workplaces," *Gender, Place and Culture* 14, no. 5 (2007): 151–67.

30. Carmen DeNavas-Walt, Bernadette D. Proctor, and Jessica C. Smith, *Income, Poverty, and Health Insurance Coverage in the United States: 2010*, P60–239 (U.S. Census Bureau, September 2011), 62.

31. Elsbeth Bösl, "Medizintechnik und/oder Lifestyleprodukt? Muttermilchpumpen, Stillsdiskurs, Konzepte von Mutter—und Elternschaft" [Medical Instrument and/or Lifestyle Product? Breastpumps, the Debate on Breast-Feeding, and Concepts of Motherhood and Parenthood], paper presented at a conference in Jena, Germany (1 June 2010); Charlotte Faircloth, *Militant Lactivism? Attachment Parenting and Intensive Motherhood in the UK and France* (Oxford, UK: Berghahn Books, 2013); and Rebecca Rosen, "A Map of Maternity Leave Policies around the World," *Atlantic* (20 June 2014), available online at http://www.theatlantic.com/business/archive/2014/06/good-job-america-a-map-of-maternity-leave-policies-around-the-world/373117/ (accessed 5 August 2014).

32. U.S. Department of Health and Human Services, Health Resources and Services Administration, Maternal and Child Health Bureau, *Child Health USA 2013* (Rockville, MD: U.S. Department of Health and Human Services, 2013).

33. U.S. Subcommittee on Health and Scientific Research of the Committee on Human Resources, 95th Congress, "First Session on Environmental Health

Sciences: Examination of the Problem of the Contamination of Mothers' Milk with Environmental Toxins and Other Problems," part 2 (Washington, DC, 8 and 10 June 1977), 7.

34. Ibid., 4.

35. Ibid., 8.

36. Margaret Munro, "Some Inuit Women Fear Breast-Milk Is Poisonous: Misguided Beliefs about Pollution Come to Light at Conference in Vancouver," *Vancouver Sun* (10 June 1995), B4.

37. M. W., "PBB Follow Up" LACTNET [Listserv] (28 May 1997), http://community.lsoft.com/scripts/wa-LSOFTDONATIONS.exe?A0=LACTNET (accessed 20 August 2014).

38. Janine DeFao, "Green Days: Bay Area Families Lead the Way in Saving the Earth," *Bay Area Parent*, East Bay edition (February 2011), available online at http://safemilk.org/wp-content/uploads/2011/02/Bay_Area_Parent_East_Bay_Edition.pdf?c99428 (accessed 3 March 2014).

39. "California Law Change Sparks Debate over Use of Flame Retardants in Furniture," *PBS NewsHour* (1 January 2014), available online at http://www.pbs.org/newshour/bb/nation-jan-june14-flame_01-01/ (accessed 3 March 2014).

Index